Synthesis Lectures on Ocean Systems Engineering

Series Editor

Nikolas Xiros, University of New Orleans, New Orleans, LA, USA

The series publishes short books on state-of-the-art research and applications in related and interdependent areas of design, construction, maintenance and operation of marine vessels and structures as well as ocean and oceanic engineering.

Alexander Arnfinn Olsen

Hazard and Risk Analysis for Organisational Safety Management

 Springer

Alexander Arnfinn Olsen
Southampton, UK

ISSN 2692-4420 ISSN 2692-4471 (electronic)
Synthesis Lectures on Ocean Systems Engineering
ISBN 978-3-031-73457-1 ISBN 978-3-031-73458-8 (eBook)
https://doi.org/10.1007/978-3-031-73458-8

This Springer imprint is published by the registered company Springer Nature Switzerland AG
The registered company address is: Gewerbestrasse 11, 6330 Cham, Switzerland

Preface

Many industries and sectors operate in hazardous and dangerous conditions. For example, the global maritime industry loses on average two vessels every day; pays out in excess $4 million in claims; and radically changes the lives of hundreds of people. Human behaviour is the source of virtually all accidents and incidents in the workplace. It is also why the loss is not significantly greater. This book is intended for non-practitioner level professionals involved in organisational safety management. It is suitable for all professionals who must be cognisant of the legal, moral, operational, and commercial demands placed on organisations to provide Safe Systems of Work (SSOW) through formal Safety Management Systems (SMS). By the end of this book, readers will be able to explain the core themes and principles of organisational safety management and apply these core principles to their work environment.

Southampton, UK   Alexander Arnfinn Olsen
February 2024

Acknowledgements It is with immense gratitude that I thank everyone at Springer for their support and engagement during the writing and publishing of this book. I would also like to acknowledge L(Phot) Alex Knott (Royal Navy) and WO Rick Brewell (Royal Navy) for providing the front cover image of Type 23 *HMS Montrose*'s Lynx Helicopter dipping her nose towards the ship during Operation Recyser (Mediterranean Sea) and RAF Nimrod MR2; both under the Open Government Licence (OGL).

Contents

Abbreviations and Acronyms

ALARP	As Low As Reasonably Practicable
CA	Criticality Analysis
CENELEC	French: Comité Européen de Normalisation Électrotechnique
	English: European Committee for Electrotechnical Standardisation
DRACAS	Data Reporting Analysis and Corrective Action System
DTI	Department for Trade and Industry (United Kingdom)
E/E/PE	Electrical/Electronic/Programmable Electronic
ESO	European Committees for Standardisation
ETA	Event Tree Analysis
EUC	Equipment Under Consideration
FHA	Failure Hazard Analysis
FMEA	Failure Modes and Effects Analysis
FMECA	Failure Mode, Effects & Criticality Analysis
FRACAS	Failure Reporting, Analysis, and Corrective Action System
FTA	Fault Tree Analysis
GAMAB	French: Globalement au moins aussi bon
	English: Globally at least as good
GAME	French: Globalement au moins équivalent
	English: Globally as least as good
HAZOPS	Hazard and Operability Study
HEART	Human Error Assessment and Reduction Techniques
HSE	Health and Safety Executive (United Kingdom)
HUD	Head Up Display
ICI	Imperial Chemical Industries (1926–2008)
ISS	International Space Station
MEM	German: Minimale Endogene Mortalität
	English: Minimum Endogenous Mortality
MOD	Ministry of Defence (United Kingdom)

MOTU   Maritime Operational Training Unit
NATO   North Atlantic Treaty Organisation
OHHA   Occupational Health Hazard Analysis
OSHA   Operating and Support Analysis
PSC   Project Safety Committee
RAF   Royal Air Force (United Kingdom)
SCP   Supplementary Conditioning Pack
SFARP   So Far As Is Reasonably Practicable
SHA   Systems Hazard Analysis
SHERPA   Systematic Human Error Reduction and Process Analysis
SIL   Safety Integrity Level
SMS   Safety Management System
SPAR(H)   Systematic Human Error Reduction and Process Analysis (Human
            Reliability Analysis)
SWIFT   Structured What If Techniques
THERP   Human Error Rate Prediction
ZHA   Zonal Hazard Analysis

List of Figures

List of Tables

Core Concepts and Themes

In this first chapter, we will begin by looking at some of the core concepts and themes you will likely encounter in your role as a maritime incident investigator. First, we will look at organisational safety management. It is important you understand what these core concepts and themes mean, as they will have a direct bearing on how you plan, prepare, carry out your investigation, interpret the root causes, and apply the principles of hazard analysis and risk management within your organisation. By organisation, we can mean either an entire company or a single operational component such as a ship. There are many different terms and words which are used in the maritime and health and safety profession, however the main terms we are interested in at this point are:

- Accident and incident
- Hazards
- Risks, and
- Safety.

It goes without saying that we all have an intuitive understanding of at least some of these concepts, but as we shall see shortly, our understanding of what these concepts mean are not necessarily be shared or universal. In other words, the term hazard may be interpreted differently from one organisation to another, and even from one person within an organisation to another person within that same organisation. For that reason, we will begin by looking at the terms "incidents and accidents".

© The Author(s), under exclusive license to Springer Nature Switzerland AG 2025
A. A. Olsen, *Hazard and Risk Analysis for Organisational Safety Management*, Synthesis Lectures on Ocean Systems Engineering, https://doi.org/10.1007/978-3-031-73458-8_1

1.1   Incidents and Accidents

The HSE, which is an independent authority of the UK Government, uses the term 'adverse event' to describe what we are calling an 'accident' or 'incident'. HSE describes an incident as "a near miss or undesired circumstance" whereas an accident is defined as an "event that results in injury or ill health". Many types of incidents and accidents are indicative of weak safety management procedures and processes within an organisation. These are often collectively called the Safety Management System (SMS). Not every organisational approach to safety management is done in the same way. Different organisations use different words and terminologies to refer to the same thing. For example, the International Electrotechnical Commission (IEC), in their standard IEC 61,508, refers to "hazardous events" instead of "accidents." The US Department of Defence (DOD), in Military Standard 882 (MIL-STD-882), refers instead to "mishaps". To complicate matters further, those organisations that do use the term "accident" frequently apply different definitions altogether. To simplify matters, throughout this textbook we will use the HSE definition of an accident, which is "an event or situation in which people are injured".[1]

1.2   Hazards

Earlier in this chapter, we across the terms "hazard" and "hazardous event". These are important terms and are in fact central to understanding the principles of organisational safety management and implementing safe systems of work (SSOW). As we saw with "accidents", there is no universally accepted definition of "hazard", though for our purposes we can turn to IEC 61,508, which defines a hazard as a "potential source of harm". In this sense, harm means any form of injury or ill health to a person or people. If harm is any form of injury or ill health caused to a person, and an accident is an event that leads to injury or ill health, we can say that a hazard is in effect a potential source of an accident. The problem with this definition is that it is too broad. This is because (a) we cannot be certain whether the potential source of an accident (the hazard) will occur; and (b) we cannot be certain we can pinpoint the exact cause of the accident. To simplify things, we can think of an organisation or process as a system. Some systems contain hundreds, and even thousands, of hazards. Some hazards are seemingly innocuous, such as standing on a stool. Other hazards have the potential to cause life changing injuries or organisational damage (for example, the sinking of the Italian cruise ship, Costa Concordia, in 2012).[2]

---

[1] HSE. 2004. Investigating accidents and incidents: A workbook for employers, unions, safety representatives and safety professionals. https://www.hse.gov.uk/pubns/hsg245.pdf.

[2] On 13 January 2012, the Costa Cruises vessel Costa Concordia (Figs. 1.1, 1.2 and 1.3) was on the last leg of a cruise around the Mediterranean Sea when she deviated from her planned route at Isola del Giglio, Tuscany. The ship was steered closer to the island and struck a rock formation on the sea floor. This caused the ship to list and then capsize, landing unevenly on an underwater

Fig. 1.1 Costa Concordia (pre-incident, 2009)

What is important to recognise is that one or more small hazards have the potential to lead to larger complex hazards. It is almost impossible for complex organisations such as shipping companies and airlines to identify every single hazard within their processes. To get around this, we can refer instead to "system level hazards".

---

ledge. Although a six-hour rescue effort brought most of the passengers ashore, 34 people died: 27 passengers, five crew, and later, two members of the salvage team. An investigation focused on shortcomings in the procedures followed by Costa Concordia's crew and the actions of her captain, Francesco Schettino, who left the ship prematurely. He left about 300 passengers onboard the sinking vessel, most of whom were rescued by helicopter or motorboats in the area. Schettino was found guilty of manslaughter and sentenced to 16 years in prison. Despite receiving its own share of criticism, Costa Cruises and its parent company, Carnival Corporation, did not face criminal charges. Costa Concordia was declared a "constructive total loss" by the cruise line's insurer, and her salvage was "one of the biggest maritime salvage operations." On 16 September 2013, the parbuckle salvage of the ship began, and by the early hours of 17 September, the ship was set upright on her underwater cradle. In July 2014, the ship was refloated using sponsons (floatation tanks) welded to her sides and was towed 200 mi (320 km) to her home port of Genoa for scrapping, which was completed in July 2017. The total cost of the disaster, including victims' compensation, refloating, towing and scrapping costs, is estimated at USD 2 billion, more than three times the USD 612 million construction cost of the ship. Costa Cruises offered compensation to passengers (to a limit of Euros 11,000; GBP 9,468; USD 11,448) per person) to pay for all damages, including the value of the cruise; one third of the survivors accepted the offer.

Fig. 1.2 Costa Concordia (post-incident)

Fig. 1.3 Costa Concordia (post-incident)

1.3 System Level Hazards

A system level hazard is a hazard that occurs on the boundary of the system in question. This may involve the failure of a system level function, or the failure of an entire system level, either of which will have an interaction with the outside world leading to an "external event". Imagine we have a railway and along that railway are a series of complex signals, junctions, and crossings. What would happen if a resistor in one of those signals failed? It might cause a red light to turn green. A train passing that signal will think it is safe to proceed when in fact there may be an obstruction on the line a mile or so around a corner. The consequences of that simple failure could be devastating. In this scenario, the resistor is a hazard, but its failure was the primary cause of the second hazard, which is the red light turning green. In essence, we have two separate hazards which, when combined, create a system level hazard. It is useful to focus on system level hazards for two reasons: (a) doing so provides a relatively tidy way of determining the chances of a hazard occurring, and the likelihood that hazard(s) will evolve into an accident; and (b) by doing so, it makes it easier to manage the overall number of hazards an organisation face. This is because the previously non-system level hazards are now considered the causes of system level hazards.

1.4 Risk and Risk Management

Risk can be defined in many ways, though the crucial components of defining risk are the frequency or probability of occurrence, and the consequences of occurrence. In other words, we can describe risk as a combination of the probability of an accident occurring and the severity of the accident should it occur. Importantly, risk is an increasing value. This means the greater the probability or severity of the accident, the greater the risk. In terms of system related risk, we refer to the combination of the risks associated with the accidents that the system can cause. Again, the risk is an increasing value. Finally, it is sometimes useful to discuss the risk of a hazard. This is defined as the combination of the probability of the hazard occurring, the probabilities of accidents resulting from the hazard, and the severity of those accidents. As we said previously, it is almost impossible for organisations to remove all hazards and risks from their systems and processes, irrespective of their severity or probability. For this reason, we turn to another concept called "tolerable risk" or "acceptable risk". In this sense, where the risk cannot be removed entirely, it is reduced to a level that is acceptable within a given context. What this means is the probability and severity of a hazard is reduced to such a level that the likelihood and consequences are tolerable or acceptable. This is the key objective of risk management. Unfortunately, determining what level of risk is tolerable or acceptable is highly subjective and depends on many interfacing factors.

1.5   Safety and Risk

Safety and risk are often interchangeable terms. For instance, if we reduce the risk of something, we make it safer. If we increase the risk of something, we make it less safe. Safety is, therefore, the absence of unacceptable risk. When we talk about safe systems, we mean the risk associated with the system is acceptable. To put this into context, we can turn to Heinrich's Triangle (Fig. 1.4). Heinrich's Triangle is a visual representation of the increasing level of risk associated with a given process—in this instance, flying an aircraft. At the bottom of the triangle, we can see there were 1,000 unreported unsafe acts. An unsafe act may be as simple as leaving a toolbox unlocked and unattended in the workshop. Above that, we can see there were ∼ 300 hazardous conditions. A hazardous condition is a situation that could have caused an incident but did not. This might include an aircraft engineer taking a spanner from the unlocked toolbox and leaving it on top of an aircraft tug. Second from the top of the triangle, we can see there were ∼ 30 incidents. An incident is the same as a near-miss, or an event that could have caused injury or ill health. Using our example, the spanner may have been knocked off the tug and left lying on the runway. At the top of the triangle, there is one aircraft accident. Although the number of occurrences has reduced as we move up the triangle, the severity and consequences of the hazards has increased quite dramatically. It is entirely feasible for the spanner that was left lying on the runway to get sucked up into an aircraft's engine causing it to crash. What seemed like an innocuous failure in safety protocol (i.e., not locking the toolbox) has resulted in a potentially devastating accident.

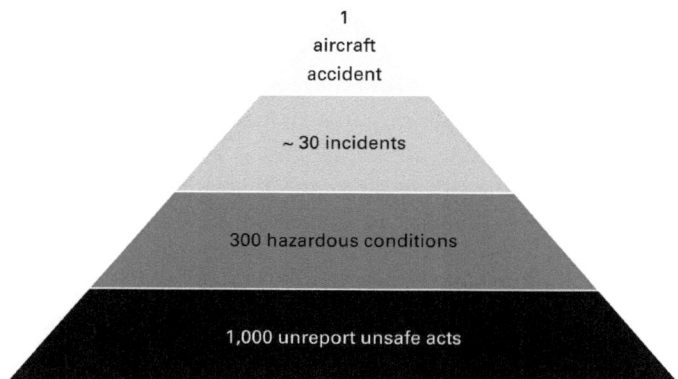

Fig. 1.4 Heinrich's triangle

1.6   Cause and Consequence

A cause is a potential event that may precipitate the occurrence of a hazard. Each cause has a probability attached to it. If we consider a fire, all fires need four elements: oxygen, heat, fuel, and a chain reaction. This is commonly referred to as the fire tetrahedron. If a piece of electrical equipment malfunctions, causing a spark, this can ignite any flammable or combustible items around it causing an electrical fire. We know the direct cause of the fire is the spark, but what caused the spark in the first place? Obviously, the electrical equipment malfunctioned, but we need to know what the initial cause was that led to the spark. In this instance, we are returning to the discussion of hazards. One small hazard (the initial malfunction) caused a larger hazard (the spark), which caused a system level hazard (the fire).

1.7   Controls and Mitigations

We have now covered some of the core themes and concepts associated with organisational safety management, which leads us to last two concepts we need to discuss: controls and mitigations. In essence, a control is a measure—be it physical or procedural—that will reduce either the probability of a cause or the probability that the cause will result in a hazard occurring. For example, by double skinning fuel pipes, we can reduce the probability that the fuel pipe will leak, or double hulling a ship, we can reduce the probability of the ship sinking if one of the two hulls are pierced. A mitigation is a form of control that limits the effects once a hazard has occurred. For example, a fire extinguisher will not prevent a fire from happening, but it will help prevent the spread of the fire when used. Alternatively, wearing a seatbelt will not stop a person from having a car accident, but it will help limit the extent of their injuries.

1.8   Swiss Cheese Model

The Swiss cheese model (Fig. 1.5) is an organisational model developed by Professor James Reason and Dante Orlandella at the University of Manchester, England. The model is used to analyse the causes of systematic failures or accidents. It is commonly used in the fields of aviation, engineering, and healthcare. The model describes accident causation as a series of events which must occur in a specific order and manner for the accident to happen. This is analogous to a series of unique slices of Swiss cheese all lined up in order. Each hole in each slice of cheese represents an opportunity for a failure to happen, and each slice represents one level in the system. A hole may allow a problem to pass through one layer, but in the next slice the holes are positioned differently. This provides an opportunity to prevent the problem from passing through to the next layer. If, however,

Fig. 1.5 Swiss cheese model

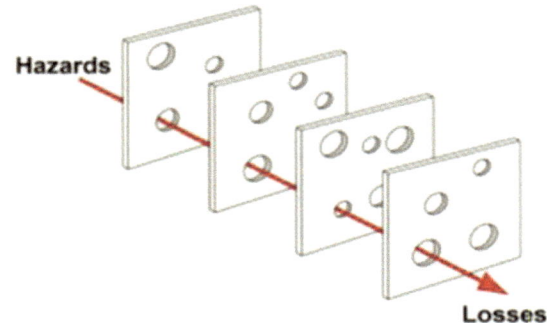

more than one slice is aligned, then the problem can freely pass from one layer of the system to the next.

In this chapter, we have been introduced to some of the basic concepts and terms used in organisational safety management. To recap on what we have covered so far:

- Safe is the absence of unacceptable risk, though we must appreciate there is an element of risk in everything we do
- Hazards, incidents, and accidents are different things
- A hazard must have the potential to result in an accident
- A cause must have the potential to contribute to a hazard
- A control must have a limiting effect on the risk, hazard, incident, or accident
- Hazards are central to system safety
- System boundary is an important element of defining hazards and mitigating risks
- It is useful to distinguish between system level hazards and causes
- Risk is a combination of the likelihood and consequences of a hazard turning into an accident.

In Chap. 2, we will turn out attention to the importance of risk in organisational safety management.

Importance of Risk in Organisational Safety Management

For most people, the idea of risk is an abstract concept. As individuals we generally know what we are prepared to do, and how far we are prepared to go, to achieve our personal aims and objectives. As humans, we are quite good at recognising our own limitations. For organisations, however, recognising risk is an entirely different proposition. The way we view a risk to ourselves is very different to how we perceive risks to our organisations. In this chapter, we will begin by reminding ourselves what a risk is, and how risks are defined at an organisational level. We will also briefly examine the historical and legal background to risk management; the perception of risk tolerability; public perceptions of risk; and the dangers of failing to take risks seriously. It is also worth reminding ourselves that when we refer to an organisation, we may refer to the company as a whole or a single operational component such as a single ship in the fleet. There are many different definitions of the word "risk", but they all tend to involve at least one of the following two components:

(1) The likelihood of something unpleasant happening, and the
(2) Consequences of something happening.

For our purposes, we can safely assume risk consists of both components, although we need not be prescriptive in how each concept is defined, except to say that increasing likelihood or severer consequences imply a greater level of risk. In terms of applying this definition of risk to an accident, we can say risk is "a combination of the likelihood of the accident occurring, and the severity of the accident". In relation to the risk of a system, we can refer instead to "a combination of the risks associated with the accidents that the system can cause". To apply these definitions, we use what is called a "risk matrix". Given

© The Author(s), under exclusive license to Springer Nature Switzerland AG 2025   9
A. A. Olsen, *Hazard and Risk Analysis for Organisational Safety Management*, Synthesis Lectures on Ocean Systems Engineering, https://doi.org/10.1007/978-3-031-73458-8_2

Fig. 2.1 MS Herald of Free Enterprise after salvage

that it is not possible to eliminate risk entirely from organisational systems, the question arises as to what is a tolerable risk? One possible answer to this question is any risk is acceptable so long as the benefit is greater than the consequence of failure. In theory, this sounds perfectly reasonable. The consequence of failing to achieve the objective is off set by the potential benefit of achieving the objective, so long, of course, as the probability of failure is within a reasonable tolerance. As you might be starting to realise, this answer is not as simple or as straightforward as it initially seems. In practice, it is often far harder to rationalise the decisions we make about risk. If we consider the sinking of the *MS Herald of Free Enterprise* in 1987 (Fig. 2.1), for example, there was always an inherent risk that the ferry might sink. The probability of the ferry sinking was considered relatively low; but the consequences were devastatingly high. For the passengers on board the ferry, the consequences of the ship sinking were offset by the low probability. Unfortunately, in this instance, the *MS Herald of Free Enterprise* did sink (due to human error). And so, it falls to safety professionals and experts to try to determine what level of risk is tolerable, whilst always remaining cognisant of public perception and opinion.

There are very few hard rules regarding the tolerability of risk. There is also little official guidance although standards such as IEC 61,508 and the UK Ministry of Defence (MOD) Safety Management Requirements for Defence Systems Part 1: Requirements (Def Stan 00–56) go some way to addressing this problem. Often, it is up to the operators, developers and vendors of safety related systems to make decisions, cast judgements, and devise protocols for determining the tolerability of risk, and in so doing, justifying those risks to sectorial regulators, the public, and when necessary, the law courts. To make matters more complicated, the tolerability of risk varies considerably across domestic, national and international borders, between industries, and even within different sectors of the same industry. When determining the tolerability of risk, it is important to consider the following factors:

(1) The absolute upper limit. Often, there is an absolute upper limit to the tolerability of risk. This however only tells us that risks are intolerable at and above this limit. It does not tell us what is tolerable or acceptable beneath this limit
(2) A comparison of new risks against existing risks. Sometimes new risks may be considered tolerable if it can be shown not to significantly increase the overall risk, or is deemed to be at or lower than the risk to be replaced
(3) The degree of control that casualties or victims have over the risk
(4) The benefits to be accrued in accepting the risk, and
(5) The practicality or cost associated with reducing the risk.

This last factor—the practicality or cost associated with reducing the risk—raises an interesting question, which is: how much is worth paying to save a single human life? It is an unpleasant question, and surely anyone should be forgiven for saying the only answer is "whatever it takes". Unfortunately, however, it is almost impossible to eliminate risk completely. Every human activity involves and requires some element of risk, whether it be flying in an aeroplane, crossing the road, or standing on a ladder. In our everyday lives we trade the benefits of the risk against the probability and consequences of the risk. For some industries, such as the maritime industry, the risks are inherently high, but it would be impractically expensive to reduce every risk associated with shipping to a level that is below our every average day. Therefore, to make a judgement about the practicality of reducing risks, it is often necessary to place a value on human life or at least on the practicality of saving a human life. Such costs should be constantly reviewed in accordance with public perception and changes to the legal framework. By current standards in the UK, the typical minimum value placed on a single human life is GBP 1.8million (USD 2,174,202; Euros 2,088,558 (November 2022). This excludes any additional compensation for injuries such as loss of sight or limbs, loss of earnings, etc. Determining tolerability of risk is a complex and emotive issue and is never easy. Fortunately, there is a method which can be used to visualise the consequences of a risk against the probability or frequency of the risk turning into an accident. This method is the risk matrix, which uses a table (shown below in Table 2.1) consisting of five levels of risk. The highest level of risk (I) is categorised as catastrophic and frequent whereas the lowest category of risk (IV) is improbable with negligible consequences. As we can clearly see from the matrix, the level of risk is the delta between the likelihood of the risk occurring and the related consequence.

Table 2.1 Example risk matrix

	Consequences			
Frequency	Catastrophic	Critical	Marginal	Negligible
Frequent	I	I	I	II
Probable	I	I	II	III
Occasional	I	II	III	III
Remote	II	III	III	IV
Improbable	III	III	IV	IV

In the next part of this chapter, we will look at perceptions of risk from an organisational perspective.

2.1   Perceptions of Risk

It is critical for organisations to never underestimate the importance and influence of public perception to risk. Today, the public are more conscious of risk than at any time in history. Social media and 24 h news broadcasting have made people increasingly aware of the world around them, and of the risks and hazards they face every day. Unfortunately, media influence has also led people to react in less rational ways. It is often argued that collectively we are getting worse at dealing with risk, and we are spending too much time and resources wrapping ourselves up in 'cotton-wool' to protect ourselves against trivial risks whilst simultaneously ignoring the larger risks. In truth, dealing with risk with any level of confidence is impossible as each new technology that evolves creates new risks. For example, although vehicle safety standards have improved in leaps and bounds over the past few decades, meaning modern vehicles are lighter and larger than ever before, this has led in part to the introduction of ancillary technologies such as satellite navigation systems, integrated media systems, and so forth. Despite modern vehicles being more structurally protected from the effects of accidents, the growth of in vehicle technology has increased the probability of accidents as drivers are more easily distracted. From the public's perspective, the convenience of having integrated vehicle technology, combined with the improvement of vehicle safety, outweighs the consequences of being involved in a potentially lethal accident.

In essence, there are many factors which influence the public's perception of risk, and ultimately, their willingness to accept risk. This means people are more likely to accept risks where they:

- Understand the technologies and dangers involved
- Recognise the dangers are distributed equitably between individuals

- Recognise individuals voluntarily take on risk
- Believe they can control their exposure to risk, and
- The consequences of any accident are immediate.

Now we have considered risk from an organisational and individual perspective, we can consider the law's point of view towards managing risk. As you might expect, the legal framework around risk management is quite complex. To summarise it, there are four key considerations that organisations must factor in relation to managing risk from a legal perspective: ethical considerations, societal considerations, commercial considerations, and legal considerations.

(1) *Ethical Considerations.* In the UK, employers have a duty of care towards their workforce. In return, workers are expected to take a commercial interest in the wellbeing of the organisation. This means working in a safe and considerate manner with due regard to health and safety, the provision and use of personal protective equipment, and protecting the organisation—insofar as is reasonably possible—from accidents occurring

(2) *Societal considerations.* Organisations have a legal duty to limit the impact of their operations, as well as their products and services, on society. This means being considerate to local communities and the environment (e.g., by not discharging waste at sea)

(3) *Commercial considerations.* Organisations have a legal responsibility to limit their exposure to the possibility of financial loss due to failure of their products and services, decreasing value of their reputation and marques, and exposure to litigation from customers, regulators and other third parties

(4) *Legal considerations.* Lastly, organisations have legal responsibilities such as the health, safety and welfare of their employees, customers and members of the public.

There is a plethora of laws, regulations, statutory instruments, and guidelines relating to the organisational management of risk. Together, these form the legal framework which governs the ways organisations plan, mitigate, manage, and respond to risks. In the UK, the number one piece of legislation that all safety professionals must be cognisant of is the Health and Safety at Work etc. Act, 1974 (HASAWA). The HASAWA was passed by Parliament into law in 1974 following the publication of the Robens Report in 1972, which recommended the formation of the Health and Safety Commission (formed in 1974 and dissolved in 2008), and the Health and Safety Executive (formed in 1975). The HASAWA places a general duty on all employers to safeguard the health and safety of their workers whilst in their employ, and sometimes even after their employment has ended. The HASAWA also places a legal duty on people who are not employees of the organisation, such as contractors, customers, and members of the public visiting the premises of the organisation; it places duties on the design, manufacture, import and export, and supply of articles and materials relating to the business of the organisation. The HASAWA also places a legal duty on the workers and employees themselves. These duties may be absolute, practicable, or reasonably practicable. Up until the mid-1980s health and safety regulation was largely prescriptive. This meant that the regulations themselves determined what was and what was not considered safe from a health and safety perspective. In 1987, following the *Piper Alpha* disaster (Figs. 2.2, 2.3 and 2.4), a new approach to health and safety was implemented in the UK following the publishing of the Cullen Report. Lord William Cullen was tasked by the government to chair the official Public Inquiry into the *Piper Alpha* disaster, which involved a fire on a North Sea oil rig some 120 miles (190 km) north-east of Aberdeen, Scotland. The incident claimed the lives of 167 men, with 30 declared missing, and over GBP 1.7billion in property and environmental damage. The ensuing Cullen Report, which ended the official enquiry in 1990 was published in two parts: the first part concerned the causes of the *Piper Alpha* disaster; and the second part made recommendations for fundamental changes to the UK's health and safety regime.

In the second part of this chapter, we have been introduced to the concept of risk, why organisations care about risk, and the legal framework around risk management. In summary, organisations have a legal duty to protect their workers, customers and the

Fig. 2.2 Piper Alpha (pre-disaster)

Fig. 2.3   Piper Alpha disaster

Fig. 2.4   Piper Alpha (post-disaster)

public from any hazards and risks arising from the organisation's operations and products. These duties are broadly defined as ethical, societal, commercial, and legal. Identifying risks and determining what is a tolerable risk is a difficult and sensitive, but necessary undertaking for safety professionals. In the next chapter, we will turn our attention towards the product life cycle and the safety life cycle and how these interact and influence an organisation's approach to safety management.

Safety Planning

To be able to investigate maritime accidents and incidents, we must first have a thorough understanding of why accidents and incidents occur in the first place. We know from the previous two chapters that safety is a legal obligation that is placed on every organisation. An accident or incident is the consequence in failing to comply with that legal duty. To ensure failures like this do not happen, organisations are required to carry out detailed safety planning in relation to their scope of operations. The planning of safety related activities which should be carried out during the development of safety related systems is a critical activity. Yet, frequently, insufficient effort and resources are spent on ensuring robust safety plans are developed, followed, and where appropriate, modified in accordance with emerging organisational needs. The safety plan should provide an initial indication of how the safety of the system is to be assured, what safety target(s) have been identified, how they will be met, and provide an outline of the strategy to be employed through which safety system objectives will be achieved and demonstrated. Like the term organisation, when we refer to system, we may refer to an individual component (such as radar), a microsystem such as a navigational watch or cargo loading operation, or a macrosystem, such as bridge or engine room operations or even the whole vessel. The first step in safety planning is to carry out preliminary hazard analysis. This is a process of identifying and qualifying potential hazards within the system.

 A. A. Olsen, *Hazard and Risk Analysis for Organisational Safety Management*, Synthesis Lectures on Ocean Systems Engineering, https://doi.org/10.1007/978-3-031-73458-8_3

3.1   Preliminary Hazard Analysis

Preliminary hazard analysis combines two key activities: initial hazard identification and initial risk assessment. The objective of the preliminary hazard analysis is to determine the safety targets for the system and the extent of risk reduction required to be implemented by the system. It is a process that is widely used to determine all requirements (i.e., safety functions and safety integrity levels (SIL). We start by dividing the hazards, accidents and the acceptable level of risk associated with each. We then identify the measures (safety requirements) to mitigate against the risks that were identified. This process requires a significant volume of work to be carried out to be effective and should never be seen or treated merely as a 'bolt on' activity. To carry out preliminary hazard analysis effectively, there are five key stages, which are outlined below:

(1) *Requirements specification.* Although an integral stage in the system lifecycle, IEC 61,508 considers requirements specification to be a standalone activity which falls outside the scope of the safety lifecycle

(2) *Design.* At the design stage, it is important to begin to allocate safety functions. This means, for example, signposting safety functions in hardware or software

(3) *Systems hazard analysis.* At this point in the safety lifecycle, we are effectively carrying out a risk assessment, but in greater and deeper detail. As we now know the design and function of the system, we can begin to estimate how the system might fail and how likely these failures are to occur. We use the same techniques as the preliminary hazard analysis, but remember, the objectives are different. Here, we want to confirm that the design meets the target level or risk

(4) *Safety validation.* For safety validation, we carry out extra testing to confirm the safety requirements of the system have been met. As we may not be able to achieve a desirable level of certainty, it may be necessary to perform additional analysis and systems modelling

(5) *Safety case report.* The safety case report is the document that summarises and pulls together all the safety activities undertaken as part of the safety lifecycle. It is used to convince Regulatory Authorities that the system is safe for operation.

3.2   Verification and Validation

Verification and validation are a combined process which crosses both safety and conventional development activities, and it is critical verification and validation activities are carried out for safety related systems. Despite its importance, there is often confusion around what verification and validation entails—even amongst published standards and guidelines. To provide some clarity on this issue, we may refer to IEC 61,508, which provides the following guidance:

(1) *Verification.* IEC 61,508 defines verification as a top down and bottom up 'V' process at each stage, where processes and procedures are appropriate and adhered to, by competent personnel, at each stage in the system development, and each safety specification complies with previous safety specifications, and there is justification for the adequacy of the tools, methods, and techniques used throughout the system lifecycle.

(2) *Validation.* IEC 61,508 defines validation as crossing the 'V', which involves simulation, analysis, testing, commissioning, product testing, and integration testing, that implements the requirements, whereby verification and validation overlaps both the safety and development lifecycles. It is important to define the strategy for achieving safety and to ensure that the verification and validation activities are sufficient to satisfy that strategy as well as demonstrating that the requirements (both functional, safety, performance, non-functional requirements) have been implemented.

3.3   Safety Planning

Commonly used safety standards such as Def Stan 00–56 and IEC 61,058 propose similar overall approaches to demonstrating safety. At the core of the process is the hazard analysis and risk assessment. Once these have been developed, appropriate measures can be incorporated into the system design. The design should be appropriately controlled, with adequate verification and validation carried out. All phases of the system lifecycle should be addressed from initial concept to end-of-life.[1] In this chapter, we will concentrate on the requirements for the plans for the management of functional safety, but also touch on the remaining aspects of planning. This will focus on the system lifecycle proposed in IEC 61,508. There are three parts to IEC 61,508: (1) System Safety Standard (IEC 61,508 Part 1); (2) Hardware Safety Standard (IEC 61,508 Part 2); and (3) Software Safety Standard (IEC 61,508 Part 3). Safety planning is an important process as it:

- Helps to define safety objectives and targets
- Helps defines the activities needed for achieving safety (i.e., for each phase of the lifecycle)
- Helps develop an understanding of the main difficulties associated with achieving the objectives, and
- Helps develop a plan for overcoming the difficulties associated with achieving the objectives.

---

[1] By system lifecycle, we are referring to the start and end of the system process. For watchkeeping, this will begin at the point the Officer/Engineer of the Watch (OOW/EOW) enters the bridge and will end when the OOW/EOW signs off the ship's log at the end of their watch.

Table 3.1 Summary of safety case composition

Risk based argument	Confidence argument	Compliance argument
Results of:	How do you do:	Compliance with:
• Hazards analysis	• Hazards analysis	• Appropriate standards and regulations
• Trials and testing	• Trials and testing	• Approved processes
• Loss modelling	• Loss modelling	
• Probability calculations	• Probability calculations	
	• Meet standards	
	• SQEP	
	• Manage risk	
	• Through life maintenance	

Often, regulatory bodies expect to see evidence of appropriate safety planning and the development of strategies for providing safety assurance. This evidence is typically compiled in a safety plan. Central to the safety plan is the safety case, which consists of three arguments: risk based, confidence based, and compliance-based arguments (see Table 3.1). There are several activities covered by the safety plan. These activities fall into three broad categories:

- *Management.* These cover the safety organisation, responsibilities, and personnel
- *Technical.* These cover safety activities and the safety life cycle, and
- *Control.* This covers the control of safety information and checking compliance and adherence to the safety plan.

Management activities include deriving safety policy and safety strategy and ensuring that all parties involved in the system development process are aware of this policy and strategy. Further management activities include determining the competencies required for carrying out each task and allocating appropriate personnel to be responsible for each activity. Technical activities cover all safety related activities within the scope of the safety lifecycle. For each activity the objective, inputs, and outputs should be clearly defined. Standards vary in their approach to documentation requirements. Some standards specify the individual documents to be produced, whereas others (including IEC 61,508) describe the material which should be documented without specifying how the material should be presented. Effective safety planning should cover both items, in the sense that the main safety documents to be produced should be described, along with some indication of the document's contents. In some cases, additional safety procedures for the control of the project may need to be further defined. For instance, procedures for managing a hazard log and for controlling the treatment of hazardous or potentially hazardous incidents will be needed in the event they do not already exist.

3.4   Allocation of Resources

The safety lifecycle can be thought of as covering three distinct phases:

- *Phase 1: Definition.* Phase 1 covered the stages from concept to safety requirements allocation
- *Phase 2: Design and Development.* Phase 2 covers the stages from overall operation and maintenance planning to demonstration of risk level acceptability, and
- *Phase 3: Operation, Maintenance, and End-of-Life (EOL).* Phase 3 covers the stages from installation and commissioning to EOL.

Typically, responsibility for safety will involve different parties at each of the different phases. The end user of the system will need to be involved during all three phases. If significant elements of the system design and development are to be subcontracted, then the subcontractors will need to be made aware of the responsibilities they are required to fulfil. Involvement of a separate maintainer in Phase 3 (a common occurrence) will produce a similar need for responsibility awareness with a separate organisation. Usually, this situation will lead to the generation of several safety plans. Each organisation will need a plan defining how its individual safety activities will be fulfilled. Taken as a whole, the safety plans should cover all activities, and describe how separate organisations will interact to ensure the safety responsibility is not falling in between organisational gaps.

3.5   Safety Plan

When preparing the safety plan, a summary should be produced to include a detailed commentary on the methods and techniques to be adopted throughout the system development lifecycle, including an evaluation of qualitative versus quantitative methods. It should also cover any relevant competency criteria, the standards to be followed, and an indication of any risk classification scheme to be adopted (together with a definition of acceptable level of risk). Once the summary has been produced and signed off, the safety plan can be developed. As we have already seen, different standards set out what they consider appropriate in terms of document contents and style. To provide some context, we will use the example provided by the UK Rail Safety and Standards Board (RSSB) Engineering Safety Management (the Yellow Book). The Yellow Book proposes the following structure:

- Safety management activities
- Safety controls
- Safety documentation
- Safety engineering, and

Table 3.2 Detailed safety plan structure

Detailed safety plan structure	
• Introduction	• Safety criteria
• Aim	• Tolerability criteria
• History of the system	• Safety requirements
• Description of the system	• Applicable standards
• Plan scope and objectives	• Standards and procedures
• Environment	• Technical plan
• System safety organisation	• Initial safety meeting
• Organisational structure	• Corporate safety culture
• Safety team objectives	• Change management
• Safety team responsibilities	• Management of trials
• Project safety team	• Incident reporting
• Engineer	• Hazard identification
• Membership	• Hazard tracking system overview
• Meetings	• Risk estimation and sentencing
• Audit plan	• Risk reduction process
• Audit process	• Verification of risk reduction
• Review process	• Safety case strategy
• Record keeping	• Safety assessment strategy

• Validation and verification of external items.

A detailed list of the safety plan contents may include any or all the following items listed in Table 3.2. To be effective, it is critical that design and safety professionals are engaged and involved from the start of the process, through each phase and stage, right up to EOL.

In this chapter, we have been introduced to the system development lifecycle and the safety lifecycle. We have seen the extraordinary amount of work that is needed to produce a well thought out safety plan. We have also begun to recognise the importance of engaging stakeholders from the beginning, and right up to the end of the system or product lifespan. In the next chapter, we will turn our attention towards preliminary hazard analysis.

Preliminary Hazard Identification and Analysis

4

In the previous chapter, we started to examine the role and function of preliminary hazard analysis, and the process of developing the safety plan. In this chapter, we will take that examination further by looking more closely at preliminary hazard identification and analysis as a function of safety planning. To begin with, we will start by looking at preliminary hazard identification.

4.1   Preliminary Hazard Identification

Preliminary hazard identification and analysis, often shortened to preliminary hazard analysis is a critical activity that is carried out early in the system lifecycle. It usually takes place before any detailed design or system development begins. There are three primary objectives to preliminary hazard identification:

(1) The identification of accidents and hazards associated with the system
(2) Analysis (often quantitative) of the ways in which accidents may develop from hazards, and
(3) Determination of system safety requirements (safety functions and associated SIL).

The word preliminary is important in this context as it not only denotes the usual place of preliminary hazard analysis within a system safety lifecycle, but also acts as an indication that the results of the analysis are often incomplete or approximate, and therefore subject to later refinement. For instance, preliminary hazard analysis tends to only identify a subset of system hazards, more of which will become apparent as the system life cycle

A. A. Olsen, *Hazard and Risk Analysis for Organisational Safety Management*, Synthesis Lectures on Ocean Systems Engineering, https://doi.org/10.1007/978-3-031-73458-8_4

develops. Preliminary hazard analysis can be split into two activities: (1) hazard and accident identification (objective 1 above) and (2) hazard and accident analysis (objective 2 above). That said, there is usually a large degree of overlap in the techniques used for carrying out the two activities. This means it is not uncommon for both activities to be performed simultaneously. For our purposes, however, we will examine both activities separately. The objective of the preliminary hazard and accident identification activity ('preliminary hazard identification') is to consider the accidents that may occur through the deployment of the system and to identify the system hazards that may lead to these accidents. Furthermore, the severities of those hazards and accidents should be identified. The preliminary hazard identification activity should attempt to determine if hazards and accidents, which have been previously identified for similar systems, could arise in the proposed system; and further establish whether the system could lead to new hazards and accidents that have not previously arisen or been identified.

The inputs to preliminary hazard identification activity should include, as a minimum, a textual description of the system and the environment in which the system will operate; and some form of diagrammatic representation of the system, including its interfaces. This representation should provide an indication of the boundary of the system, together with checklists covering generic hazards that have been identified in similar systems. The outputs of the preliminary hazard identification activity should include, as a minimum, preliminary accident lists, and a preliminary hazard list. The preliminary hazard list will tend to indicate the system hazards in very general terms. For example, a preliminary hazard list for a head-up display (HUD) on a radar terminal might include incorrect data showing on the screen; no information showing on the screen; delayed display of information on the screen, or a confusing or muddled display of information on the screen. It is recommended that preliminary identification is a team-based activity; ideally conducted through one or more round table discussions or meetings. The team should include domain experts, safety engineers, designers, and end user representatives. The team should agree the scope of the system and determine whether hazards that have previously arisen in other similar systems could arise in the proposed system. One or more models of the system should be used in the meetings to stimulate thought and discussion. The model representations should provide an overview of the system and its functions including the system interfaces, for example:

- System inputs
- System outputs
- Interfaces with other systems, and
- Human interfaces.

In addition to identifying system interfaces, the analysis should also examine the system boundaries as these are provide determining differences between the system operation, hazards and failures resulting in accidents. Moreover, the analysis should identify and describe the environment in which the system is to be used. This is important since a critical hazard in one environment may not be so critical in another (for example, the use of an automated air conditioning system is likely to be more critical in a ship's engine room than in a private car); and be supported by textual descriptions of the system functionality which should include the modes of operation of the system (for example, start-up, initialisation; normal operation; and shut down). Techniques that can be used to perform preliminary hazard identification include group round table discussions based on agreed system model(s), checklists of questions to be posed to experts individually and or in group meetings, hazard and accident data collation from similar systems, litera-ture searches, relevant database scrapes, etc., and a formal Hazard and Operability Study (HAZOP) (which is also used for analysis of hazards and accidents). Table 4.1 describes some of the advantages and limitations of these techniques.

During preliminary hazard identification, the following problems may be encountered:

(1) Getting the level of detail within the system representation right is difficult (too much detail and the identification process can get bogged down in detail; too little detail, and there is insufficient information available to identify hazards and accidents
(2) There is a tendency to spend (too much) time and resource designing the system rather than identifying hazards
(3) Failure to consider maintenance and operational issues, and
(4) It can be difficult to control group activities whilst maintaining a systematic (complete) approach to avoid unnecessary digression.

Once the preliminary hazard identification stage is complete, we can move onto the next stage which is preliminary hazard and accident analysis.

4.2   Preliminary Hazard and Accident Analysis

The objectives of preliminary hazard and accident analysis (preliminary hazard analysis) or preliminary hazard analysis is to establish the relationship between system hazards and the accidents they can cause; determine system safety requirements; continue the process of hazard and accident identification; and begin to identify the causes of accidents. The inputs to the preliminary hazard analysis should include (1) a preliminary hazard and accident list arising from the hazard identification activity; (2) a briefing document, delivered to all team members, prior to any team meeting activities; this briefing document should set out the objective of the preliminary hazard analysis, the scope of the system, the functionality of the system, one or more diagrammatic representations of the system, and a definition of the technique(s) to be used for the analysis activity; to include a definition of the roles of each attendee. The outputs from the preliminary hazard analysis should be used to initiate a hazard log. The hazard log is an evolving document which should

Table 4.1 Advantages and limitations of preliminary hazard analysis

Technique	Advantages	Limitations
Round table discussion	Encourages innovative thought so may identify potential hazards and accidents that have not occurred in the past but may occur in future; involves a combination of safety engineers, the customer, and designers to ensure the widest possible scope of experience and knowledge is utilised; less time consuming and less rigorous, which means more detailed structural analysis techniques such as HAZOP can be used for low-risk systems	Unstructured, so can easily miss hazards and accidents; without a formal structure, it is not easy to define the scope of the analysis and there is a tendency to consider details beyond the system under consideration; can be time consuming as lacking in formal structure; can be difficult to document as meetings are unstructured and discussions may become 'side-tracked'; unlikely to provide sufficient rigour for higher risk systems as discussions are not systematic and are prone to incompleteness
Checklist	Encourages consideration of all aspects of the system, including operation, maintenance, and decommissioning (provided the check lists cover these aspects); can be combined with a round table discussion to structure the session and make use of the advantages of the round table discussion approach; different check lists can be used to encourage consideration of many aspects of the proposed system	Does not fully support innovative thought; there may be a tendency to assume that use of the checklist is sufficiently robust to identify hazards; checklists need to be kept current to ensure they retain currency; Checklists need to be customised to ensure they are relevant to the system under consideration

(continued)

Table 4.1 (continued)

Technique	Advantages	Limitations
HAZOP	Provides a structured analysis technique by application of guide words to each major component of the system; use of guide words to steer the analysis promotes a more innovative and application specific focus compared to the checklist approach; involves a combination of safety engineers and the customer to ensure that wide experience and knowledge is utilised; there is a large body of experience in the application of HAZOPs which is widely documented; used to determine the consequences of a hazard; further consideration is given to the possible mitigations that may already exist in the design	Insufficient information for complete analysis since a clear understanding of the system is required; with insufficient detail, numerous assumptions may be made; all assumptions must be verified; any erroneous assumptions will affect the validity of the HAZOP activity; time consuming and may have to be repeated when more information is available; needs a representation of the system which can be understood by all parties involved; needs guide words which are appropriate for the system under analysis; it can be difficult to schedule meetings involving the appropriate personnel; the analysis can become side-tracked unless meetings are controlled by a chairman experienced in running HAZOPS; it can be difficult to put together the ideal team of SQEP individuals
Functional hazard analysis	Used to determine whether failures in functions will lead to the hazards identified; a systematic technique; helps to identify which sub-systems are likely to be critical; can be conducted by individuals working alone	Requires a thorough understanding of the system functions which may not be available at this stage of the development; will not utilise a wide range of expertise if conducted in isolation by only one individual
Accident sequences and event tree analysis	Used to provide a diagrammatic representation of the combination of events or circumstances which must occur for a hazard to lead to an accident; supports quantitative reasoning about the likelihood of hazards leading to accidents	Tendency to assume that a sequence of events or circumstances must be in place for a hazard to lead to an accident, which may not be true

record all the identified system hazards together with information about the level of risk associated with each hazard. The preliminary hazard analysis activity should consider the behaviour of the system and whether this behaviour could lead to any of the hazards identified in the preliminary hazard list, or indeed, any hazards not previously identified. Furthermore, the relationship between the hazards and accidents should be considered in detail including the system modes or operational scenarios in which a hazard can credibly lead to an accident; the likelihood of being in those system modes or operational scenarios; the events or circumstances that must pertain for the hazard to lead to an accident; and the likelihood of those events or circumstances occurring. The techniques that can be used to perform preliminary hazard analysis include group based round table discussions on agreed system model(s); checklists of questions to be posed to experts individually and or in group meetings; a formal HAZOP, Functional Hazard Analysis (FHA), Accident Sequences, and Event Trees.

- **HAZOP**. The Hazard and Operability Study or HAZOP is a formal process in which a group of experts with designated roles (including the chairperson, domain experts, system designers, system users and secretary) meet to systematically consider all system interfaces, the methods with which they can fail, and the consequences of failure. They will identify hazards and accidents, determine the relationship between them and, to a limited extent, identify potential causes of the hazards and possible mitigations
- **FHA**. Functional Hazard Analysis or FHA is a technique through which the functions of a system are systematically analysed, whilst considering the consequences of possible failures of those functions. It can be used to identify system hazards and accidents, and, to a limited extent, the causes of hazards
- Accident Sequence. An accident sequence is the progression of events that can result in an accident. It is particularly useful for modelling the sequence of events that must occur, or not occur, for a hazard to lead to an accident
- **Event Trees**. Event trees are a particular diagrammatic representation of accident sequences that facilitate quantitative analysis of how likely it is that a hazard will lead to an accident. Table 4.1 describes the advantages and limitations of these techniques.

During the preliminary hazard analysis, the following difficulties may be encountered: in general, it is hard to determine the events or circumstances which must occur for a hazard to lead to an accident. Moreover, it is hard to quantify the likelihood of those events or circumstances occurring. This is because there is a tendency to design the system rather than identify the hazards and issues around the system. It can be difficult to obtain sufficient commitment from the customer and designers to the analysis activity. The result is an insufficient consideration of all aspects of the system and the environment in which the system will operate.

4.3   Hazard and Operability Studies (HAZOP)

The HAZOP was first developed by the chemical industry for determining if the failure of components in a process plant could lead to hazards. The HAZOP is based on a structured round table discussion, where guide words to each component within a system are applied to identify deviations from intent. Judgement and experience are then used to determine whether these deviations could lead to hazards within the system.

4.4   Guide Words

The guide words to be used for the HAZOP should be chosen to reflect the components under analysis. To be effective, the guide words need to be pertinent to the component. If ambiguous or loose words are chosen, time and effort can be wasted attempting to interpret the meaning and context of the component being analysed. Attributes are assigned to each component, which are then supported by the guide words selected. Table 4.2 provides some examples of commonly used guide words.

In this chapter, we have looked at the theory behind preliminary hazard analysis. We have examined hazard and accident identification, which is based on previous experience, checklists of questions, etc., and a systematic study of system interfaces. We have also examined hazard and accident analysis techniques, which involves the identification of accident sequences, mitigating events, and so forth. Lastly, we looked at hazard and operability studies which is the systemic study of system interfaces by a formal group of domain experts. In the next chapter, we will look at the concept of functional safety.

Table 4.2 Example guide words

Component	Attribute	Guideword	Interpretation
Data	Flow	No More Part of Reverse	No informational flow More data is passed than expected The information passed is complete Flow of information in wrong direction
	Rate	More Less	The data rate is too high The data rate is too low
	Value	More Less	The data value is too high The data value is too low
State transition diagrams	Event	No As well as Part of Other than	Event does not happen Another event take place as well Only some of the needed conditions occur An unexpected event occurs instead of the anticipated event
	Action	No As well as Part of Other than	Action does not take place Additional (unwanted) actions take place An incomplete action is performed An incorrect action takes place
Time	Timing	No Early Late Before/after	The event/action never takes place The event/action takes place before it is expected The event/action takes place after it was expected Happens out of expected order of event/action
	Repetition	No More Less As well as Other than	Output is not updated Time between outputs is longer than required Time between outputs is shorter than required Synchronisation with other repetitions causes problems Time between outputs is variable

Functional Safety

In this chapter, we will be introduced to the process of identifying safety functions and deriving associated SIL. Collectively, these are referred to as the safety requirements for a system with the potential to give rise to accidents. For clarity, we can turn to IEC 61,508, Part 4, which provides us with the following definitions:

(1) **Safety function**—"[a] function to be implemented by an E/E/PE safety related system, other technology safety related system[s] or external risk reduction facilities, which is intended to achieve or maintain a safe state for the EUC [equipment under control], in respect of a specific hazardous event"
(2) **Safety integrity level**—"[the] discrete level (one out of a possible four) for specifying the safety integrity requirements of the safety functions to be allocated to the E/E/PE safety related system[s], where safety integrity 4 has the highest level of safety integrity and safety integrity 1 has the lowest"
(3) **Safety integrity**—"[the] probability of a safety related system[s] satisfactorily performing the required safety functions under all stated conditions within a stated period of time".

At the level of detail considered during initial hazard and risk analysis activities, the list of potential examples of safety functions are wide and varied, but may include:

- Automotive—such as throttle control, brake control, crumple zones, etc.
- Process control—such as emergency shut down, fire and gas detection, and explosion containment barriers.

31
A. A. Olsen, *Hazard and Risk Analysis for Organisational Safety Management*, Synthesis Lectures on Ocean Systems Engineering, https://doi.org/10.1007/978-3-031-73458-8_5

At the level of detail considered during the design of subsystems, for instance programmable electronic equipment, it would be usual to specify safety functions using logic or mathematics-based notations and in terms of the relationship between specific inputs (causes) and outputs (effects). A distinction is often made between 'control functions' and 'protection function'. In truth, either may be classed as a safety function though it is important to understand the distinction as there are often subtle differences in the way they are handled within industry standards. For example, IEC 61,508 places restrictions on when 'control functions' may be considered a 'safety function'.

5.1   Types of Safety Function

For our purposes, there are two main types of safety function: control functions and protection functions. These are both discussed below.

5.2   Control Functions

Control functions are those functions, performed by the system, which respond to input signals causing the system to operate in the desired manner. These might include, for example, functions of an aircraft flight control system. In this scenario, the aircraft flight control system causes the aircraft to respond in the desired manner to signals from the pilot, and to other aircraft control process signals. By way of comparison, a chemical process control system is also a control function in the sense that it causes the various pumps and valves to respond in the desired manner in accordance with instructions received from the operator as well as in response to level, flow and other process control feedback signals.

5.3   Protection Functions

Protection functions are those functions that are intended to prevent the system from behaving in some undesirable or unintended manner. Where the undesirable behaviour of the system leads to a hazardous situation and could therefore lead to an accident occurring, the protection functions are also considered to be safety functions. Examples of protection functions that would be performed by a programmable electronic system might include an alarm, cut-off, or shutdown function. These functions are performed within separate and dedicated equipment, for instance, a separate fire and gas protection system, or an emergency shutdown system.

5.4   Safety Integrity Levels (SIL)

In the previous section, we noted how safety functions are those functions that impact the risk of a potential accident occurring within an associated system. Given that the system(s) performing the safety functions may fail, the risk of an identified accident also depends on the probability that the safety functions are performed correctly when required. This is recognised within many industry standards (and IEC 61,508 in particular) by requiring that—in addition to specifying the safety functions that affect the risk of an identified accident—the necessary safety integrity of each safety function should also be specified. In practice, the necessary safety integrity of a safety function is specified in terms of the probability that the safety function will be performed correctly when required, or inversely, that the safety function will not be performed correctly (i.e., fail) when required (on demand). For safety functions (e.g., control functions) that are required continuously, the probability is expressed in terms of failure within a specified period. Where safety functions are to be performed by electrical, electronic, or programmable electronic equipment, many standards (particularly IEC 61,508) requires that the necessary safety integrity of each safety function is specified in terms of a safety integrity level. For this purpose, Tables 2 and 3 in Part 1 of IEC 61,508 relate probability of failure directly to one of the four safety integrity levels. The specified safety integrity level is used by IEC 61,508 as a means of establishing four best-practice benchmarks; one for each safety integrity level, against which faults within the safety-related system(s), and its development process, can be compared, prevented (at best), or tolerated (at worst). It should be noted that there is no universal acceptance of the concept of safety integrity levels and the associated best-practice benchmarks provided within IEC 61,508. Some practitioners argue the benchmarks provided by IEC 61,508 are a useful guide to establishing whether a particular system and its development process are adequate for a particular application. Others argue, however, that technology develops too quickly and the differences between different applications are too great to be able to establish universally applicable benchmarks.

5.5   SIL Probabilities

IEC 61,508 provides parameters for what is considered the tolerable probability for one dangerous failure per X demands or Y hours of operation (Table 6). Alternatively, we can refer to the standards in Tables 5.1 and 5.2.

Table 5.1 SIL probabilities

SIL 1	SIL 2	SIL 3	SIL 4
Probability of dangerous failures per hour of between 10–5 to 10–6	Probability of dangerous failures per hour of between 10–6 to 10–7	Probability of dangerous failures per hour of between 10–7 to 10–8	Probability of dangerous failures per hour of between 10–8 to 10–9
Once every 10 demands	Once every 100 demands	Once every 1,000 demands	Once every 10,000 demands
Like SIL 2 but with lower number of design techniques to apply	Requiring good design and operating practice (like ISO 9001), with appropriate design techniques	Less onerous than SIL 4 but still requiring high level of design effort/operating and training practice	Highest target and the most onerous, requiring state of the art techniques

Table 5.2 Variations in SIL probabilities (1)

Standard	Measure (derivation)	High	Low	N/A
IEC 615,508	Safety integrity level (from system level)	4	1	–
CENELEC 50,129	Software Safety Integrity Level (from system level 50,126/50,129)	4	1	0
Def Stan 00–55	Safety Integrity Level (from system level 00–56)	4	1	1
MIL-STD-882C	Software Hazard Risk Index	1	5	1
RTCA DO 178C	Software Level (from system level consequences)	A	D	E

5.6   Risk Reduction Process

There are three steps related to the risk reduction process. The first step in the process towards identifying safety functions and deriving associated SIL is to identify the set of possible accidents associated with the system under consideration. We do this by assuming that any known (control or protection) functions of the system could fail. For each accident, the acceptable risk associated with the accident is also identified and specified. The second step in the process is to designate known functions of the system as safety functions on the basis that they contribute to the prevention of one or more of the identified accidents. In doing so, a decision needs to be made as to whether the control system of the overall system under consideration is itself to be designated as a safety related system. Here, the approach adopted within different industry sectors varies. For instance, in the aviation industry, it would be highly unlikely for the control system (e.g., the flight control system) not to be designated a safety related system as it is difficult to conceive a separate protection system. Within the process control industry, however, it is normal

Table 5.3 Variations in SIL probabilities (2)

Standard	Measure (derivation)	High	
IEC 615,508	Safety Integrity Level (from system level)	4	Probability of dangerous failure in range of ≥ 10–9 to 10–8 per hour for continuous mode of operation
CENELEC 50,129	Software safety integrity level (from system level 50,126/50,129)	4	Tolerable hazard rate of ≥ 10–9 to 10–8 per hour per function
Def Stan 00–55	Safety integrity level (from system level 00–56)	4	Minimum failure rate is "remote"—i.e., "likely to occur sometime"
MIL-STD-882C	Software hazard risk Index	1	High risk—significant analysis and testing resource
RTCA DO 178C	Software level (from system level consequences)	A	Failure condition classification of catastrophic

practice to consider the process control system as being separate and distinct from the safety related system. This is because in the process industry, safety of a process is normally assured using separate protection systems that operate independently of the control system providing some degree of redundancy. In the event of abnormal operation, the separate protection system forces the process into some form of predefined safe state. In the process industry it is normally possible to force systems into safe states; stating the obvious, this is not normally possible in aviation except, perhaps, by (crash) landing the aircraft. The third step in the process is to estimate the frequency (or probability) of occurrence of each of the identified accidents in the absence (i.e., total failure) of the designated safety functions. This is the EUC risk as shown at Figure A.1, Part 5, of IEC 61,508 (Table 5.3).

This frequency (or probability) may be extremely high (always) if the normal control functions of the system have themselves been designated as safety functions. The contributions made by each safety function to the prevention of a specific accident is then considered. Each safety function is considered to 'reduce the risk' associated with the accident and the combination of the risk reduction performed by each safety function is the 'actual risk reduction' shown at Figure A.1, Part 5, of IEC 61,508. The combination of the 'actual risk reduction' with the 'EUC risk' leads to a 'residual risk' associated with the accident. The residual risk is compared with the specified acceptable risk and, if necessary, additional (or different) safety functions are identified and or the safety integrity requirements of the safety functions are altered. It is worth noting that there may not be a unique allocation of safety integrity levels that achieves the required risk reduction. Having identified a set of safety functions, and their safety integrity requirements, which adequately reduce the risk associated with a specific accident, the safety functions are allocated to the various systems that will perform the functions. Where safety functions are allocated to electrical, electronic, or programmable electronic equipment, the safety integrity requirements are converted to safety integrity levels using the two tables in Part 1 of IEC 61,508.

5.7   First Principles (Quantitative) Approach

From the overview of the risk reduction process, which we covered in the previous section, it should be fairly evident that the risk associated with a specific accident is reduced (i.e., managed) through a potentially complex combination of safety functions with different safety integrity requirements attached to them. To identify the required (target) safety integrity of each safety function, it is necessary to be able to model the interaction of the various designated safety functions (control and protection), other failure conditions of the system and external parameters, such as exposure time. The required probabilities of failure of the designated safety functions can then be combined to provide the probability of occurrence of the accident of interest (the residual risk). A technique such as fault tree analysis may be used to model the interaction of the designated safety functions and to combine the required probabilities of failure of the safety functions. In IEC 61,508, Part 5, Annex C, an example is presented of the calculation of the required safety integrity in terms of the average probability of failure on demand (PFD_{avg}) of an individual safety function. The formula used by IEC 61,508 depends on the acceptable frequency of the accident of interest (F_a) and the predicted frequency of the accident in the absence of the safety function of interest (F_{np}). Having identified the required safety integrity of a safety function in this way, the required safety integrity level to be attached to the safety function is simply derived from IEC 61,508, Part 1, Table 2.

As an illustrative example, given an acceptable frequency of once every 400 years (0.0025 per year) for a specific accident and an estimate that the accident would occur once every 2 years (0.5 per year) in the absence of a designated safety function, the required safety integrity (probability of failure on demand) of the designated safety function is $PFD_{avg} < Fa/F_{np} = 0.0025/0.5 = 0.005$. In accordance with Table 2, Part 1 of IEC 61,508, this equates to a required SIL of SIL 2. In practice, the task of evaluating the combination of probabilities is more difficult than the simplistic example given in IEC 61,508, Part 5, Annex C. As such, some of the most salient points to consider include the provision of care needed to be given when combining probabilities relating to failure within a given period with probabilities relating to failure on demand. This can only work where a commonly defined period is used throughout the calculation. Other factors such as the level of exposure to the risk can have a significant effect on the calculation and need to be carefully allowed for; and believable data needs to be obtained on the different initiating events (failures in the absence of safety functions) that could lead to an accident before an attempt can be made to measure the required risk reduction. For example, when

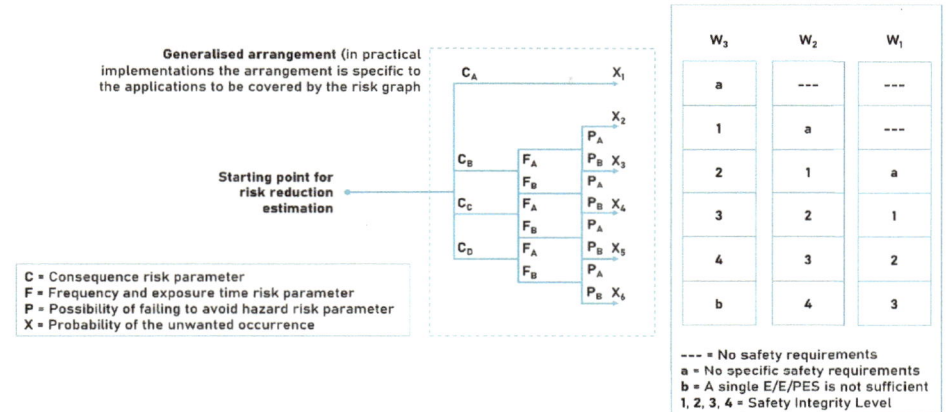

Fig. 5.1   Risk graph (adapted from IEC 61,508)

considering the need for, and required safety integrity of, a gas detection system, a thorough analysis needs to have been carried out of the modes and frequencies of failure of the process plant that could lead to dangerous gases being released into the atmosphere.

5.8   Other (Qualitative Approaches)

Because of the effort required to calculate the required safety integrity of a safety function from first (quantitative) principles, different industry sectors have formulated their own approaches which consider specific factors of the industry or application sector and hide some of the complexity of the calculations. However, these approaches need to be carefully calibrated before they can be used. Two such approaches are illustrated in IEC 61,508, Part 5. Annex D presents an illustration of the risk graph approach; Annex E presents an illustration of the hazardous event severity matrix approach. The diagrams from IEC 61,508 Part 5 of are shown below as Figs. 5.1 and 5.2.

5.9   Applying the SIL

The specified SIL dictates the way in which a safety related system is developed to implement its specified safety function(s). A fuller description of how the required safety integrity level influences the development of a safety related system is given in the subsequent chapters of part 1 of this textbook. To that end, suffice it to say that IEC 61,508 requires that, where multiple safety functions of different required SIL are to be implemented in a single system, the highest SIL should apply to the development of the whole

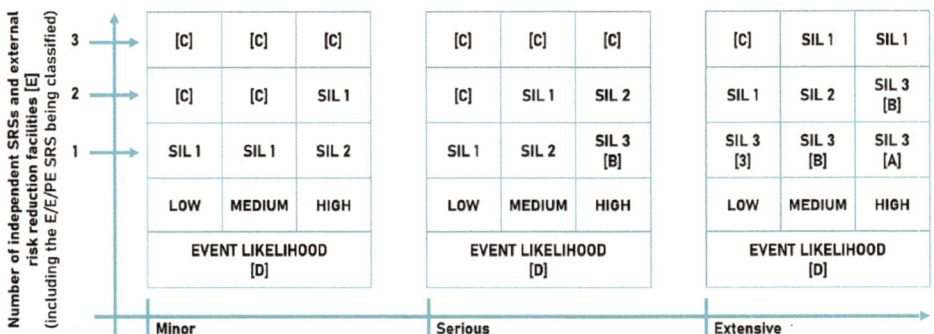

Fig. 5.2 Hazardous event severity matrix (adapted from IEC 61,508)

system unless adequate separation within the system can be demonstrated. Where a safety related system incorporates some form of software in the design of that system, the SIL that applies to the system, also applies to the development of the software for the system in question. In Parts 2 and 3 of IEC 61,508 (requirements for developing the system and its software) two new concepts are introduced: (a) random hardware safety integrity and (b) systematic safety integrity (also sometimes termed software safety integrity). Random hardware safety integrity is designed to represent the component of the required safety integrity relating to random hardware failures of the safety related system, for example component wear-out failures. Systematic safety integrity is designed to represent the component of the required safety integrity relating to systematic failures of the safety related system arising from, for example, design faults or limitations. With respect to hardware safety integrity, the system designer is required to demonstrate that the probability of component hardware failures leading to failure of the system to correctly perform the safety function when required is no higher than the required safety integrity (probability of failure) of the safety function. This is done through consideration of factors such as redundancy within the system architecture, diagnostic coverage, diagnostic test intervals and proof test intervals. With respect to systematic safety integrity, the system designer is required to incorporate features within the design of the system (and its software) that would tolerate the effects of systematic failures (failures arising from errors made by the designer) and to use techniques and measures that help to prevent such systematic failures (prevent errors being made and or overlooked during the development process). In Parts 2 and 3 of IEC 61,508, tables are presented that relate different design features, and development techniques and measures to the four SILs.

In this section, we will briefly discuss the views from industry on the concepts of safety functions and SIL. These views were captured by the UK Government during a Department for Trade and Industry (DTI) sponsored research project on the practical application of IEC 61,508. Although IEC 61,508 clearly relates SIL to safety functions, SIL are very often used in other contexts, with a SIL being quoted with respect to a system, an item of equipment of even a component used within an item of equipment. The consensus of industry experts regarding this issue is as follows:

- In accordance with the intention of IEC 61,508, SIL should be identified with the safety functions to which they apply
- Safety functions are themselves composed of a hierarchy of subfunctions (lower-level safety functions). SIL may need to be identified with the lower-level safety functions depending on the level in the hierarchy of subfunctions at which the risk assessment/ reduction process has been applied
- There is wide use of the SIL as a measure of the capability of COTS equipment, for example programmable logic controllers intended for safety related applications. Although not ideal, this situation will need to be accepted. It is essential, however, that suppliers identify the type of function(s) to which it is assumed the equipment will be applied and to specify all constraints (including environmental) that apply
- SIL should not be identified in relation to components as they will normally be used within a wide range of different systems.

IEC 61,508 defines failure probability targets for each of its four SIL [IEC 61,508, Part 1, Tables 2 and 3]. The standard then requires that the identified failure probability target be applied to both random hardware failures and systematic failures associated with the system used to implement the required safety function(s) [IEC 61,508, Part 2, Clause 7.4.3, Note 1]. Although conventional reliability theory allows predictions to be made of the reliability of a system's hardware components, there is not a similar, accepted, theory that can be applied to the design process. Recognising this, IEC 61,508 provides derived criteria against which to demonstrate compliance (the tables of recommended techniques and measures). Therefore, it should not be necessary for developers to have to predict explicitly the systematic failure probability of a safety related system.

The SIL, when viewed simply as a target (index), enables developers to discriminate between the importance (criticality) of different functions; to divide up available resources based on the importance (criticality) of different functions; to focus resources on the most critical functions; and to make decisions based on the rigour of evidence to be provided within the safety case. In Part 5 of IEC 61,508, guidance is also provided on the methods of determining the required SIL for identified safety functions. The examples illustrated in Part 5 show how a SIL is determined for safety functions that are identified as a means of reacting to a hazardous situation and thus avoiding an accident. However, it is not clear, from the guidance provided in Part 5, how the SIL can be determined when it is the safety function itself (e.g., a ship's steering function) that gives rise to the hazardous condition. Whilst it is recognised the presentation of IEC 61,508 is clearly biased towards the process industry sector and 'protection' type safety functions, the principles of risk assessment/risk reduction apply equally to both 'protection' type safety functions and 'control' type safety functions.

5.10   Targets for Hardware Failure Probability

IEC 61,508 requires that the SIL should be determined by comparing the risk of an identified accident without the provision of a particular safety function and the acceptable risk of the identified accident scenario; the SIL is a component of the necessary risk reduction. Having identified the SIL, it is then applied equally with respect to random hardware failures and systematic failures of the system implementing the safety function. The problem with this approach is that it does not allow for the fact that the acceptable risk of a particular accident scenario may, itself, be a function of the cause of the accident: one value if the cause is component wear out (random hardware failure); another value if the cause is human error during development or during operation (systematic failure). Theoretically, there is no reason why the hazard and risk analysis process described in Part 1 of IEC 61,508 could not be applied separately in relation to accidents caused by random hardware failures and accidents caused by systematic failures.

Part 2 of IEC 61,508 already separates safety integrity into systematic safety integrity and random hardware safety integrity. The point at which the separation occurs could simply be moved to the start of the hazard and risk analysis process rather than the end of the hazard and risk analysis process. It is worth noting that Def Stan 00–56 only uses the concept of SIL with respect to systematic failures.

5.11   Relationship Between Safety Integrity Levels, Techniques and Measures

IEC 61,508 presents a correspondence between recommended techniques and measures to be used to avoid or tolerate the introduction of systematic failures within a safety-related system design and the four SIL. However, there appears to be little (if any) published evidence to support the correspondences made within the standard. The tables of techniques and measures presented within IEC 61,508 are useful in so far as they provide a 'cookbook' of ideas for organisations not already well advanced in safety related system development processes. However, a detailed rationale, supported by evidence, should be developed to provide credence to the correspondence claimed within IEC 61,508 between techniques, measures and safety integrity. Further guidance should also be provided on required levels of achievement associated with the use of the identified techniques and measures. For example, achievement against defined test coverage metrics could be identified for each SIL.

In this chapter, we have looked at some of the key concepts and principles around safety requirements. Having identified the potential accidents that could occur because of operating some process or equipment, what actions should we take? The answer is we need to put in place measures to prevent the identified accidents from occurring and decide how effective those measures need to be. The 'what' and 'how effective' are referred to as safety requirements and need to be specified carefully if we are going to develop and implement a process or system that is safe. In the next chapter, we will start to look at risk analysis, first by discussing what risk is and how it is defined, and then by examining methods of analysing system level risks.

Understanding Risk Analysis

The aim of this chapter is to explain the 'risk-based approach' commonly adopted for safety related systems, and to introduce techniques for risk estimation. The relationship of risk assessment to other activities in the safety lifecycle will be discussed together with a critique of the limitations associated with the risk analysis model. It is worth pointing out at this initial stage that all systems have some degree of risk. To establish ways of reducing that risk, and to make claim that the level of risk is acceptable, we must first understand the nature of the risk involved. There are three universally accepted models for assessing risk. These are:

- ALARP—As Low as Reasonably Practicable (UK)
- GAMAB (Globalement au moins aussi bon—generally at least as good) (France)
- MEM (Minimale endogene Mortalität—minimum endogenous mortality) (Germany).

6.1 ALARP (As Low As Reasonably Practicable)

As we noted earlier, in the UK (and across the world) there is a legal requirement to demonstrate that the risks (to health and safety) associated with a system or process have been identified and understood. Since all systems have a degree of risk, it goes without saying that at least some element of risk must be tolerated. Hence, in the UK, the concept of as low as reasonably possible or ALARP is used to link the level of risk to the cost/practicality of reducing the risk to an acceptably tolerable level. The concept of ALARP originates from the HASAWA, which requires the provision and maintenance of systems which are designed and built to be operated in such a way that they are safe

A. A. Olsen, *Hazard and Risk Analysis for Organisational Safety Management*, Synthesis Lectures on Ocean Systems Engineering, https://doi.org/10.1007/978-3-031-73458-8_6

and without health risks "so far as is reasonably possible" (SFARP). The definition of SFARP in this context carries the requirement that risks must be reduced to a level that is at a minimum ALARP. When determining whether a risk is ALARP, it is necessary to define what "reasonably possible" means. This term has formed a part of English law since Edwards vs. The National Coal Board in 1949. In that ruling, it was decided that the risk must be significant in relation to the sacrifice (i.e., money, time, and inconvenience) necessary to avoid the risk. In other words, the risk must be avoided unless the difference between the cost and the benefit obtained is 'disproportionate'. By including the term "disproportionate", determining whether a risk is ALARP is not as straightforward as say a simple cost-benefit analysis, as the balance will (always) be tilted towards making safety improvements. That said, there is no universally accepted consensus on what defines an acceptable or appropriate measure. Therefore, in the context of ALARP, risk is defined as the combination of frequency (probability) and the consequence (importance or severity) of an accident or hazardous event.

The factors which are generally considered when determining whether a risk is tolerable (in the context of ALARP) include:

- Safety and hygiene regulations
- Specifications
- Current regulations and legislation, and
- Guidance from advisory bodies.

Provided the risk has been previously reduced to a tolerable level, to determine whether the risk has been reduced to the ALARP level, a cost-benefit analysis must be carried out. This analysis should demonstrate, through comparison, that the cost required to reduce the risk would be disproportionate to the benefit obtained by reducing the risk. In the event this cannot be demonstrated, the residual risk cannot be considered ALARP. Another factor to consider is the cost of evaluating the benefit obtained in risk reduction. In extremely complex systems, this cost can be prohibitively expensive, and may in fact be the determining factor when considering the feasibility of risk reduction. The ALARP triangle or 'carrot diagram' is often used to illustrate the types and levels of risks within the concept of ALARP. It is called a carrot diagram because of its triangular shape, with the apex at the bottom, and the frequent use of different shades of orange (the universal colour for hazards). The diagram indicates at the top the unacceptable risks which must be reduced without considering the economic benefit of reducing the risk. At the bottom of the diagram are those risks considered insignificant. The intermediate area is the zone of tolerable risk, also called the ALARP Zone, though this is inaccurate as the principle of ALARP is applicable to all levels.

6.2   GAMAB (Globalement au Moins Aussi Bon)

The French model Globalement au moins aussi bon (GAMAB) or Globalement au moins équivalent (GAME) dictates that any new system should be at least as safe and low risk as any existing comparable system. This principle sets the current level of safety as a minimum requirement. In other words, any new system must be as good as or better than the existing safety level. Unlike the ALARP model, which considers risks on an individual level, GAMAB requires that the sum of individual risks must not exceed the total risk level of the existing system. Similarly, the total risks of any new system must not be higher than those already present in the current system with comparable performance characteristics and operating conditions. The consideration of total-risk provides some room for sensible risk trade-offs; a system may be considered acceptable even if the risk of one individual aspect is increased, provided the overall total-risk level is maintained by overcompensating for risks elsewhere in the system. GAMAB supports rationale decision-making by placing the focus on the totality of the system and provides a clear baseline that is not typically open to the same level of interpretation like the subjective requirements of the ALARP principle. Moreover, GAMAB does not require decision-makers to prove the cost-effectiveness of a safety system as is the case with ALARP. This provides decision-makers with greater latitude when designing and implementing risk-based safety systems (Fig. 6.1).

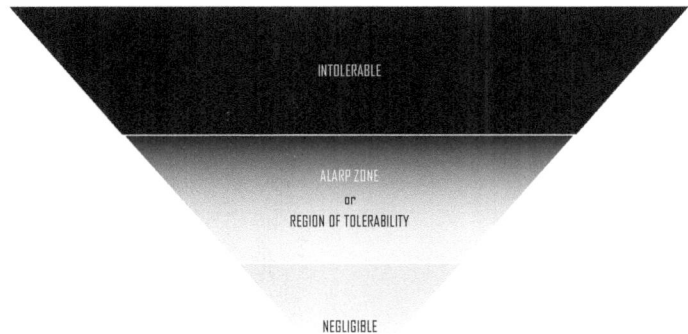

Fig. 6.1 ALARP triangle

6.3   MEM (Minimale Endogene Mortalität)

The third model, or the German minimale endogene Mortalität (minimum endogenous mortality (MEM) is a measure of the accepted (inevitable) risk of death from the system or process. It is described in the CENELEC standard EN 50126, which stipulates the acceptable safety level is a maximum of 0.0002 deaths per person per year. Statistically speaking, this the mortality (risk of death) of an average European adolescent. MEM is usually used as an absolute risk threshold for the approval of complete systems. New systems must not have a higher risk than the existing system (the same principle is used in GAMAB). Since everyone is exposed to 'many' (standardised: 20) technical systems at any given time, a threshold of 1/20 MEM or 0.00001 deaths per year is generally set for each individual system. This value must not be exceeded by planned innovations. On the contrary, new technologies must be held safer than the system to be replaced. This is taken from the British ALARP principle. It is important to recognise that MEM has some limitations. First, it is not certain that a system that meets the MEM criterion will also meet the European Committee for Standardisation (ESO) standard. Second, as MEM sets a fixed risk criterion, it is absolute and cannot easily adapt to societal changes. The perception of whether something is too risky largely depends on numerous factors; for example, whether there is an alternative available, or whether there have been recent accidents that have raised public awareness.

6.4   Conceptualising Risk

Risk arises from accidents, also called 'hazardous events' or 'mishaps'. The measure of risk is a combination of the severity of the accident and the frequency or probability with which the accident occurs. It is important to use a consistent measure of either frequency or probability.

6.5   Risk Estimation

There are two main approaches to risk estimation:

(1) Qualitative, and
(2) Quantitative (numerical).

With risk estimation, several bands of frequency and severity are defined, with the combinations of frequency and severity divided into classes of risk. The highest class of risk is unacceptable; the lowest class of risk is acceptable. In between the two levels are two intermediate classes of risk. This area is the ALARP region (Tables 6.1, 6.2, 6.3, 6.4 and 6.5).

In 1972 the Ford Motor Company introduced a new model of passenger car, which they called the Ford Pinto (Fig. 6.3). The Pinto had no reinforcement between the pliable rear bumper, the fuel tank and the differential on the rear axle. In the event of a rear shunt, the fuel tank would be pushed onto the rear differential, rupturing the fuel tank causing

Table 6.1 Accident frequency categories (Def Stan 00-56)

Accident frequency	Occurrence during the operational life considering all aspects of the system
Frequent	Likely to be continually experienced
Probable	Likely to occur often
Occasional	Likely to occur several times
Remote	Likely to occur some time
Improbable	Unlikely, but may exceptionally occur
Incredible	Extremely unlikely that the event will occur at all, given the assumptions recorded about the domain and the system

Table 6.2 Accident frequency categories (rail and nuclear industry (UK))

Accident frequency	Nuclear industry	Rail industry
Frequent	$10{,}000 \times 10^{-6}$/operating hour	Daily to monthly
Probable	100×10^{-6}/operating hour	Monthly to yearly
Occasional	1×10^{-6}/operating hour	Yearly to 10 yearly
Remote	0.01×10^{-6}/operating hour	10 yearly to 100 years
Improbable	0.0001×10^{-6}/operating hour	More than 100 years
Incredible	$0.000001 \times 10\text{-}6$/operating hour	—

Table 6.3 Risk matrix (UK defence, rail and nuclear sectors)

Frequency	Consequence			
	Catastrophic	Critical	Marginal	Negligible
Frequent	24	18	12	6
Probable	20	15	10	5
Occasional	16	12	8	4
Remote	12	9	6	3
Improbable	8	6	4	2
Incredible	4	3	2	1

Table 6.4 Risk matrix (non-sector specific industrial)

	Consequence			
Frequency	Catastrophic	Critical	Marginal	Negligible
Frequent	I	I	I	II
Probable	I	I	II	III
Occasional	I	II	III	III
Remote	II	III	III	IV
Improbable	III	III	IV	IV
Incredible	III	IV	IV	IV

Table 6.5 Hazard severity classifications

Category	Definition
Catastrophic	Multiple fatalities
Critical	One or two fatalities; and or multiple severe injuries; and or severe occupational illnesses
Marginal	A single severe injury; and or multiple minor injuries; or minor occupational illnesses
Negligible	At most a single minor injury or minor occupational illnesses

fuel to ignite. The lack of reinforcement on the doors also meant they were likely to jam shut by the same force resonating from the rear. This led to the Ford Pinto gaining the unfortunate moniker, "Ford's barbeque that seats 4". In response, Ford carried out a cost benefit analysis that looked at the feasibility of recalling the Pinto and retrofitting adequate rear and side impact protection. The analysis concluded it would be cheaper to pay out compensation for each death and injury than pay the approximate USD 11 per vehicle needed to improve the vehicle's safety (at the time of the analysis it was estimated there were 2.2 million registered Pinto's on US roads). However, in 1981 an incident involving the 13-year-old Richard Grimshaw,[1] almost cost the Ford Motor Company USD 125million in compensation and punitive damages after the teenager sustained life threatening injuries and burns following a rear shunt at only 28mph (45kmh). The jury in the case decided that the Ford Motor Company should be penalised to such an extent that any compensation awarded would be equal to the entire profit Ford made on selling the Pinto up to that point. By 1981, this was estimated at approximately USD 125million. On appeal, the actual award for compensation and punitive damages was reduced to USD 3.5million; a sizeable sum for 1981 but considerably less than what the jury wanted to award. During the trial, it was alleged that hundreds of deaths had resulted from the Pinto's poor design

---

[1] Grimshaw vs. Ford Motor Company (119 Cal.App.3d 757, 174 Cal.Rptr. 348).

Fig. 6.2 Titanic Sinking, engraving by Willy Stöwer (1912)

and Ford's refusal to remediate those vehicles already purchased though later research showed that the actual number of deaths was around 28. Had the original award stood, it would have bankrupted Ford, both in terms of the punitive damages but also through loss of reputation. This then begs an obvious question: should the Ford Motor Company have introduced the risk reduction measure? From an emotive perspective, the answer of course must be a resounding 'yes'. But what did the cost benefit analysis that Ford carried out show? By comparing both sides of the ledger, it is perfectly clear why Ford decided not to retrofit the Pinto with risk reduction measures: from a financial point of view, it was more economical to leave the vehicles as they were than to retrofit each vehicle (see Table 6.6).

6.6   Frequency and Sequencing

The frequency of an accident depends on the frequency with which the hazard leading up to the accident in question arises and the probability of this hazard leading to the accident. A carefully considered definition of the way the accidents can occur is therefore needed. The hazard may only arise in certain circumstances; for example, when someone is within a restricted area. Often, there are other events which must occur for the hazard to lead to an accident. The sequence of events through which a hazard leads to an accident is known as the accident sequence. A quantitative or numerical risk assessment requires a numerical based measure of accidents, such as cost (i.e., money) or number

Fig. 6.3 Ford Pinto (1973)

Table 6.6 Cost benefit comparison (Ford Pinto, 1972 USD)

Cost of not retrofitting the Ford Pinto		Cost of retrofitting the Ford Pinto	
180 fatalities	$200,000 per fatality	11 million cars	$11 per vehicle
180 serious burn casualties	$67,000 per injury	1.5 million light trucks	$11 per vehicle
2,100 affected vehicles	$700 per vehicle		
Total estimated benefit: $49.5 million		Total estimated cost: $137 million	
(180 × $200,000) + (180 × $67,000) + (2,700 × $700)		(11,000,000 × $11) + (1,500,000 × $11)	

of fatalities. Other events, such as injuries and non-fatal casualties, may then align to the chosen measure. The frequency of each accident is calculated using fault tree or event tree analysis using numerical probabilities. Quantitative or numerical risk analysis is attractive when sufficient quality data is available (for instance, through operational experience). In practice, however, quantitative, and qualitative methods are usually combined. It is not uncommon for the quantitative analysis of the accident sequence to be combined with the use of a risk matrix to define acceptable risk. From this, it is easier to 'close out' hazards (or accidents) one by one. By 'close out', we mean the hazard is judged to be acceptable.

6.7   Safety Targets

The risk matrix is a form of safety target. All hazards must have an acceptable class of risk assigned to it. This means we can then assess the safety target in terms of the chosen measure of accidents – for example, by the number of fatalities per year or fatalities per passenger mile. Alternatively, safety targets can be comparative – for example, the system is as safe as before. Distinct safety targets may be agreed for different groups of people who may be harmed by an accident involving the system. For instance, one safety target may be defined for trained (e.g., engineering) personnel who are directly involved with the system; another, more stringent, safety target may be defined for people, such as deck officers, who are indirectly involved with the system during their duties; and a third, more stringent still, safety target may be defined for people who are independent third parties, such as deck ratings or passengers.

6.8   Risk Classification

The various standards and industry guidelines have different ways of demonstrating risk classification. For example, IEC 61508 uses a matrix where each square has a derived classification number in Roman numerals and a corresponding colour. Alternatively, several industry standards use the 5 x 5 matrix. To calculate the risk, the frequency and consequence rating are multiplied. If we recall that the ALARP principle is to keep risks 'as low as reasonably possible', we can reference the IEC 61508 classifications of risk against ALARP. This provides us with a clear demarcation of risk. Risk analysis, especially quantitative risk analysis, is often the subject of criticism. The main issues with risk analysis is the difficulty associated with estimating the frequency of occurrence accurately. This is especially so when the probability of an event occurring is unlikely. For example, in the Leveson Report, it was suggested that the probability of being hit by a falling aeroplane is between 10-7 and 10-8 during an average person's lifetime.[2] Subsequently, calculating a frequency of 10-6 per hour for this type of accident means that the occurrence cannot arise from the analysed causes. In general, the correctness of the numbers arising from the quantitative risk assessment is dependent on the accuracy of the system model generated for calculation purposes. Frequently, this model will incorporate approximations and estimates with large error margins, leading to results with similar characteristics. To assess a given risk, we need to know how likely accidents are to occur, and how often they are likely to occur. We also need to consider the anticipated consequences should the accident occur. We can only anticipate the consequences as accidents will have likely effects (which we can determine with a large degree of certainty) but also unlikely effects (which are possible, but unlikely).

---

[2] In the UK, that is 82 years for the average male. By comparison, the chance of winning the lottery is estimated at 1 in 45,057,474 out of a total population of 67.33million.

To put this another way, let us take the example of a train derailment. When a train derails (i.e., comes off the tracks) it is very likely that the train crew and passengers will sustain injury, but – depending on the severity of the accident – it may or may not be likely that fatalities are involved. This of course depends largely on the contingent circumstances around why the train derailed in the first place. For instance, at Great Heck in North Yorkshire, England, a high-speed train collided with a vehicle that had broken down on the East Coast Main Line on 28 February 2001. Although there were initially no fatalities, this changed when an oncoming goods train crashed into the immobile passenger train, killing 10 and injuring 82. What seemed at first to be a critical accident very quickly escalated into a catastrophe. When assessing risk severity, therefore, it is equally important to consider the peripheral factors as it is to consider the primary factors. Taking the train derailment as a case in point, there are several direct consequences that we need to consider. First, we can combine the two first train derailment with the secondary collision. Although separate incidents, one is directly interlinked with the other. The peripheral factors are the sub sequent clean-up of debris, repairs that must be made to the tracks and railway infrastructure, and compensation to be paid to the passengers who sustained injuries. We can also safely assume there were knock on losses caused by the closure of the line as the trains were removed and the tracks fixed. For train operators, these are risks that are inherent and must be accepted as a hazard of the industry. As individual consumers, however, we must also accept a degree of risk, which is defined by ALARP as individual risk. This relates to the probability of harm to the individual. Ironically, train travel is substantially safer than travelling by car, but as individuals, we often make the subconscious decision to accept the higher risk of driving than to accept the substantially lesser risk of train travel. Often this decision is made because of comfort and convenience rather than by any conscious estimation of risk.[3]

6.9   Categories of Severity

Severity is often categorised into sub-levels, which allows for easier risk analysis and interpretation. These categories often differ from one sector to another, which means there is no universally accepted definition for severity. For example, IEC 61508 merely states "each accident severity shall be categorised during Risk Estimation in accordance with the definitions of Table 2 but if these definitions are not appropriate for the system being considered they may be modified to include other aspects such as system loss

---

[3] Statistically speaking, in 2011 there were 373 fatalities on commercial passenger flights compared to 2.84billion commercial passengers. This means the odds of dying on a commercial flight are approximately one in 7.6million. The World Health Organisation estimates that the chances of dying in a vehicular accident in the US is one in 83 or equal to 1.2million deaths per year worldwide. By comparison, between 2005 and 2011 only 16 people died in cruise accidents out of a global total of 100million passengers, which is roughly equal to one in 6.25million over five years.

or environmental effects. Any modifications shall be agreed by the Independent Safety Auditor and the accident severity categories used for the system shall be recorded in the Hazard Log". Def Stan 00-56, on the other hand, helpfully provides us with the following severity categories:

Accordingly, there are generally two options for reducing risk: (a) reduce the likelihood of occurrence; and (b) reduce the severity. The ways of implementing either or both options are largely dependent on the nature of the risk, the process or system in question, and the domain in which the process or system exists. Risk reduction is a tri- tier process consisting of three measures: (1) the design stage (e.g., introducing redundancy into the system; adding extra functions; etc.); (2) the operational stage (e.g., training for specific activities); and (3) the maintenance stage (e.g., preventative, and corrective maintenance; frequency of parts replacement; etc.) A pertinent example of this is the RMS Titanic (1912) (Fig. 6.2).[4] Although the ship was fitted out with lifeboats (the protective measure) there were insufficient lifeboats available for the entire crew and passenger count. This inevitably led to unnecessary loss of life. Today, ships must carry sufficient lifesaving appliances (lifeboats, life rafts and life vests) for the maximum number of crew and passengers onboard. Although the ship may still be lost, in theory the number of fatalities resulting from the loss of the ship should be considerably lower. Sometimes it is possible to reduce the frequency of an accident by designing out failures in the system that have led to accidents occurring in the past. This involves placing barriers which prevent the hazard from being the cause of future accidents. Although it is not possible to remove the risk entirely, by reducing the likelihood of the hazard, it is possible to lower the overall risk; bearing in mind, of course, that if the accident occurs, it may still be catastrophic (for example, an air crash). Reducing the consequence of the accident involves the provision of protection against the anticipated effect (consequences) of the accident. In many situations this is extremely difficult or costly to achieve, but should the accident occur, the likely consequences should not be as catastrophic compared to the protective measures not being in place.

6.10   Case Study: Ford Pinto (1972)

In this chapter we have looked at some of the key themes and concepts concerning risk. We now know how to define risk, how to estimate risk, and how to define the severity of hazards before and after they turn into accidents. We have also looked at the various ways of reducing risk, and why it is important to manage risk responsibly, if not necessarily

---

[4] RMS Titanic was a British passenger liner, operated by the White Star Line, which sank in the North Atlantic Ocean on 15 April 1912 after striking an iceberg during her maiden voyage from Southampton, England, to New York City, United States. Of the estimated 2,224 passengers and crew aboard, more than 1,500 died, making it the deadliest sinking of a single ship up to that time. It remains the deadliest peacetime sinking of a superliner or cruise ship.

ethically. In the next chapter of this textbook, we will examine several risk assessment methodologies such as systems hazard analysis, fault tree analysis, failure modes and effects analysis and bow tie analysis.

The first step in assessing risk is to carry out a process called Systems Hazard Analysis (SHA). The results or output of the SHA is the identification of different types of hazards. As we have said, a hazard is any potential condition that either exists or it does not exist (i.e., probability of 1 or 0). It may (in single or in combination with other hazards (events) and conditions become an actual functional failure or accident (mishap). The sequence in which this series of occurrences happen is called the scenario. The scenario has a probability (i.e., between 1 and 0) of occurrence. Some systems and processes may have many potential failure scenarios. It is also assigned a classification based on the worst-case severity of the end condition. Risk is the combination of probability and severity. Preliminary risk levels can be provided through SHA. The validation, more precise prediction (i.e., verification), and acceptance of risk is determined in the risk analysis (risk assessment). The primary objective of both analyses is to provide the best selection of means available for controlling or eliminating the risk. SHA is employed in several sectors including avionics, chemical process safety, safety engineering, reliability engineering, and food safety. A hazard is defined as a "condition, event, or circumstance that could lead to—or contribute to—an unplanned or undesirable event". It is very rare for a single hazard to cause an accident or a functional failure. Rather, it is by far more common for accidents and operational failures to be the result of a sequence of causes. SHA considers the system state (for example, the operating environment), together with the opportunities for potential failures and malfunctions. Whilst in some cases safety and reliability risk can be eliminated, in most situations a certain degree of risk will always remain and must be accepted. To quantify expected costs before the fact, the potential consequences and the probability of occurrence must be considered. Assessment of risk

© The Author(s), under exclusive license to Springer Nature Switzerland AG 2025   55
A. A. Olsen, *Hazard and Risk Analysis for Organisational Safety Management*, Synthesis
Lectures on Ocean Systems Engineering, https://doi.org/10.1007/978-3-031-73458-8_7

is carried out by combining the severity of the consequences with the likelihood of occurrence using the severity matrix. Risks that fall into the "unacceptable" category (i.e., high severity with high probability) must be mitigated by some means to reduce the level of safety risk. For example, IEEE STD-1228–1994 'Software Safety Plans'[1] prescribes industry best practices for conducting software safety hazard analyses to help ensure safety requirements and attributes are properly defined and specified for inclusion in software that commands, controls or monitors critical functions. The severity of consequence identified by the SHA establishes the criticality level of the software. Software criticality levels range from Alpha (A) to Echo (E), corresponding with to the severity of Catastrophic to No Safety Effect. Higher levels of rigour are required for level Alpha and Bravo software and their corresponding functional tasks and work products.

7.1   Systems Hazard Analysis (SHA)

There are three primary types of SHA. The first is Operating and Support Analysis (OSA). This type of systems hazard analysis examines the hazardous tasks undertaken by operational and support personnel. The second is Occupational Health Hazard Analysis (OHHA). This type of SHA identifies threats to health from hazardous substances, temperatures, electricity, radiation, etc. The third is Zonal Hazard Analysis (ZHA) which examines potential interactions between systems in relation to their physical locations within the system environment. For ships and aircraft, ZHA is particularly relevant as evidenced by the 2006 crash of the RAF Nimrod XV230 in Afghanistan.[2] The circumstances surrounding the XV230 incident is discussed in further detail in the next chapter.

7.2   Functional Systems Hazard Analysis (FSHA)

There are three types of FSHA:

(1) Fault Tree Analysis (FTA)
(2) Event Tree Analysis (ETA), and
(3) Failure Modes Effects Analysis (FMEA) and or Failure Modes Effects and Criticality Analysis (FMECA).

---

[1] https://ieeexplore.ieee.org/document/9097571.

[2] On 2 September 2006, a Royal Air Force Hawker Siddeley Nimrod suffered an inflight fire and subsequently crashed in Kandahar, Afghanistan, killing all 14 crew members on board. The crash, which occurred during a reconnaissance flight, was the biggest single loss of life suffered by the British military since the Falklands War.

When carrying out FSHA, there are several potential data sources which can be mined, including historic failure data (including specific failure databases (e.g., Failure Reporting, Analysis and Corrective Action System (FRACAS) and Data Reporting, Analysis and Corrective Action System (DRACAS); published failure data; and generic failure prediction techniques (e.g., MIL-HD-BK-217F); and human error calculations (including human error assessment and reduction techniques (HEART); Technique for Human Error Rate Prediction (THERP); Systematic Human Error Reduction and Process Analysis (SHERPA); and SPAR(H) Human Reliability Analysis.

7.3   Fault Tree Analysis (FTA)

FTA is a top-down, deductive failure analysis process through which undesired states of a system can be analysed using Boolean logic to combine a series of lower-level events. This analysis is mainly used in safety engineering and reliability engineering to understand how systems can fail; to identify the best way of reducing risk; and to determine event rates of a safety accident or a particular system level (functional) failure. FTA is also extensively used in software engineering for debugging purposes and is closely related to the cause-elimination techniques used for detecting software bugs. In the aerospace sector, the general term "system failure condition" is used instead of "undesired state" or top event of the fault tree. These conditions are classified in accordance with the severity of their effects. The most severe conditions require the most extensive FTA. These system failure conditions, and their classification are often determined through functional hazard analysis. FTA provides a useful tool for understanding the logic leading to the top event/undesired state; showing compliance with the (input) system safety and reliability requirements; prioritising the contributors leading to the top event (i.e., creating the critical equipment, parts and events lists for different important measures); monitoring and controlling the safety performance of complex systems (for example, whether a particular aircraft safe to fly when fuel valve x malfunctions? Or how long is the aircraft safe to fly with the fuel valve x malfunction?); minimising and optimising resources; assisting in the design of a system (for example, the FTA can be used as a design tool that helps to create (output/lower level) requirements; and functioning as a diagnostic tool to identify and correct causes of the top event, which can then be used in the creation of diagnostic manuals and processes.

FTA methodology is described in various industry and government standards including NRC NUREG-0492 for the nuclear power industry, an aerospace-orientated revision to NUREG-0492 which is used predominantly by NASA, SAE ARP 4761 for civilian aerospace, MIL-HDBK-338 for military and defence systems. IEC 61,025 is intended for cross-sector use and has been adopted as European Norm EN 61,025. Any sufficiently complex system is subject to failure because of one or more subsystems failing. The likelihood of the system failure, however, can often be reduced through the imposition of improved system design. Fault tree analysis maps the relationship between faults, sub-systems, and redundant safety design elements by creating a logic diagram of the overall system. The undesired outcome is taken as the root ('top event') of a tree of logic. For example, the undesired outcome of a metal cutting process is the removal of the oper-ator's hand. Working backwards from this top event, we might determine there are two possible ways this might happen: the first is during normal operation, and the second is during maintenance. This condition is a logical OR. Considering the branch of occurring during normal operation, we might further determine there are two opportunities where the operator might suffer injury: the saw cycles and harms the operator; or the saw cycles and harms another person. This is another logical OR. In response, we can make a design improvement by requiring the operator to press two buttons to cycle the machine. This is a safety procedure in the form of a logical AND. The button may have an intrinsic failure rate—this becomes a fault stimulus that we can analyse. When a fault tree is label with numbers for failure probabilities, analytical software can calculate the actual failure probabilities from the fault tree. When a specific event is found to have more than one effect event, this is called a common cause or common mode. On a diagram, we can see how common modes appear at several locations in the tree. Common modes introduce dependency relations between events. The tree is usually written out using conventional logic gate symbols.

Many different approaches can be used to model an FTA, though the most common way is to (1) define the undesired event to analyse (definition of the undesired fault can be difficult to uncover, though many faults are obvious. An experienced engineer with a wide knowledge of the system is best placed to help define and number the undesired events; obtain and understanding of the system to be analysed; once the undesired event has been determined, all causes with probabilities of affecting the undesired event of 0 or more are studied and analysed. Producing exact numbers for the probabilities leading to an event is usually impossible as such deep level analysis is time consuming and costly. To aid in this process, computer software may be used; system analysts may be able to pro- vide a wider understanding of the system. System designers will have a full knowledge of the system and this knowledge is critical to ensuring no potential causes affecting the undesired event are missed; for the selected event all causes are then numbered and sequenced in the order of occurrence; these are then used for the next step which is drawing or constructing the fault tree; (2) construct the fault tree (after selecting the undesired event, and having analysed the system to the extent that we know all the causing effects (and if possible, their probabilities) we can now construct the fault tree. The fault tree is based on 'AND' and 'OR' gates which define the major characteristics of the fault tree; (3) evaluate the fault tree (after the fault tree has been assembled for a specific undesired event, it is evaluated and analysed for any possible improvement (or in other words, study the risk management, and find ways for system improvement. A wide range of qualitative and quantitative analysis methods can be applied); (4) control the hazards identified (this step is very specific and will necessarily differ from one system to another).

FTA uses symbols to represent key stages in the system process. These symbols are grouped together as events, gates, and transfer symbols. Although the symbols used in FTA are standard, minor variations may be found in different software platforms. Event symbols are used for primary events and intermediate events. Primary events are not developed further on the fault tree. Intermediate events are found at the output of a gate. The primary event symbols are typically used as follows:

- *Basic events*—failure or error in a system component or element (e.g., a switch stuck in the open position)
- *External events*—normally expected to occur (but not of itself a fault)
- *Undeveloped event*—an event about which insufficient information is available, or which is of no consequence
- *Conditioning event*—conditions that restrict or affect logic gates (e.g., the mode of operation in effect)
- An *intermediate event* gate can be used to immediately above a primary event to provide more room to type the event description.

Alternatively, gate symbols describe the relationship between the input and the output events. The symbols are derived from Boolean logic symbols.

7.4   FTA Software

In the US, the most common FTA software programme is the Electric Power Research Institute's CAFTA software,[3] which is used by many US nuclear power stations and by a majority of American and overseas aerospace manufacturers. Additionally, Idaho National Laboratory's SAPHIRE[4] is used extensively by the US Government to evaluate the safety and reliability of nuclear reactors, the Space Shuttle, and the International Space Station (ISS). Outside the US, RiskSpectrum[5] is used by approximately half of the world's non-US nuclear power stations for probabilistic safety assessment. Alternatively, some organisations use the open-source software SCRAM, which implements the Open-PSA Model Exchange Format open standard for probabilistic safety assessment applications. Some industries use both FTA and ETA. We will explore ETA in the next section, but it is worth explaining here by way of introduction that the event tree starts from an undesired event (the initiator, such as a loss of critical supply or component failure) and follows possible further system events through to a series of final consequences. As each new event is considered, a new node on the tree is added with a split of possibilities of taking

---

[3] https://www.epri.com/research/products/000000003002000020.

[4] https://saphire.inl.gov/#/.

[5] https://www.riskspectrum.com/.

either branch. From this point, the probabilities of a range of 'top events' arising from the initial event can be determined.

7.5   Event Tree Analysis (ETA)

ETA is a form of forward, top-down, logical modelling that is used for both success and failure analysis. It explores responses through a single initiating event and lays a path for assessing probabilities of the outcomes and overall system analysis. This analysis technique is used to analyse the effects of functioning or failed systems given that an event has occurred. ETA is a powerful tool that identifies the consequences of a system that have a probability of occurring after an initiating event that can be applied to a wide range of systems, including nuclear power stations, spacecraft, and chemical processing plants. This technique may be applied to a system early in the design process to identify potential issues that may arise, rather than correcting the issues once they have occurred. With this forward logic process, the use of ETA as a tool in risk analysis can help prevent negative outcomes from occurring, by providing a risk assessor with the probability of occurrence. ETA uses a type of modelling technique called the event tree, which branches events from one single event using Boolean logic. Performing a probabilistic risk assessment starts with a set of initiating events that change the state or configuration of the system. An initiating event is an event that starts a reaction, such as the way a spark (the initiating event) can start a fire that could lead to other events (intermediate events) such as a tree burning down, and then finally an outcome, for example, the burnt tree no longer providing apples for food. Each initiating event leads to another event, and continuing through this path, where each intermediate event's probability of occurrence may be calculated by using fault tree analysis, until an end state is reached (the outcome of a tree no longer providing apples for food). Intermediate events are commonly split into a binary (success/failure or yes/no) outcome but may be split into more than two if the events are mutually exclusive. This means that they cannot occur at the same time. If a spark is the initiating event there is a probability that the spark will start a fire or will not start a fire (binary 'yes' or 'no') as well as the probability that the fire spreads to a tree or does not spread to a tree. End states are classified that can be successes or severity of consequences.

An example of a success would be that no fire started, and the tree still provided apples for food while the severity of consequence would be that a fire did start, and we lose apples as a source of food. Loss end states can be any state at the end of the pathway that is a negative outcome of the initiating event. The loss end state is highly dependent upon the system, for example if you were measuring a quality process in a factory a loss or end state would be that the product must be reworked or thrown in the rubbish. Some common loss end states include:

- Loss of life or injury or illness to personnel
- Damage to or loss of equipment or property (including software)
- Unexpected or collateral damage as a result or tests
- Failure of a mission or process
- Loss of system availability, and
- Damage to the environment.

The overall objective of ETA is to determine the probability of possible negative out-comes that can cause harm and result from the chosen initiating event. It is necessary to use detailed information about a system to understand intermediate events, accident scenarios, and initiating events to construct the event tree diagram. The event tree begins with the initiating event where consequences of this event follow in a binary (success/failure) manner. Each event creates a path in which a series of successes or failures will occur where the overall probability of occurrence for that path can be calculated. The probabilities of failures for intermediate events can be calculated using fault tree analysis and the probability of success can be calculated from $1 =$ probability of success (ps) + probability of failure (pf). For example, in the Eq. $1 = (ps) + (pf)$ if we know that pf $= 0.1$ from FTA then through algebra we can solve for ps where ps $= (1)-(pf)$ then we would have ps $= (1)-(0.1)$ and ps $= 0.9$. The event tree diagram models all possible path-ways from the initiating event. The initiating event starts at the left side as a horizontal line that branches vertically. The vertical branch is representative of the success/failure of the initiating event. At the end of the vertical branch a horizontal line is drawn at the top and the bottom representing the success or failure of the first event where a description (usually success or failure) is written with a tag that represents the path such as 1 s where s is a success and 1 is the event number similarly with 1f where 1 is the event number and f denotes a failure. This process continues until the end state is reached. When the event tree diagram has reached the end state for all pathways the outcome probability equation is written. The steps to perform an event tree analysis include:

- **Define the system**: define what needs to be involved or where to draw the boundaries
- **Identify the accident scenarios**: perform a system assessment to find hazards or accident scenarios within the system design
- **Identify the initiating events**: use a hazard analysis to define initiating events
- **Identify intermediate events**: identify countermeasures associated with the specific scenario
- **Build the event tree diagram**
- **Obtain event failure probabilities**: if the failure probability cannot be obtained use fault tree analysis to calculate it
- **Identify the outcome risk:** calculate the overall probability of the event paths and determine the risk

- **Evaluate the outcome risk:** evaluate the risk of each path and determine its acceptability
- **Recommend corrective action:** if the outcome risk of a path is not acceptable develop design changes that change the risk
- **Document the ETA:** document the entire process on the event tree diagrams and update for new information as needed.

7.6   Advantages and Limitations of ETA

The main advantages of using ETA are it enables the assessment of multiple, co-existing faults and failures; it functions simultaneously in cases of failure and success; there is no need to anticipate end events; the areas of single point failure, system vulnerability, and low payoff countermeasures may be identified and assessed to deploy resources; the paths in a system that lead to a failure can be identified and traced to display ineffective countermeasures; work can be computerised using COTS software; it can be performed on various levels of details; there is a visual cause and effect relationship; it is relatively easy to learn and execute; it models' complex systems into an understandable manner; it follows fault paths across system boundaries; it combines hardware, software, environment, and human interaction into one; and it enables probability assessments. On the other hand, the limitations of ETA are it only addresses one initiating event at a time; the initiating challenge must be identified by the analyst; the pathways must be identified by the analyst; the level of loss for each pathway may not be distinguishable without further analysis; success or failure probabilities are difficult to establish; it is easy to overlook subtle system differences; partial successes and failures are not distinguishable; and ETA requires an analyst with practical training and experience.

7.6.1   FMEA

FMEA is the process of reviewing as many components, assemblies, and subsystems as possible to identify potential failure modes in a system together with their causes and effects. For each component, the failure modes, and their resulting effects on the rest of the system are recorded in a specific failure mode and effects analysis worksheet (Fig. 7.2). There are numerous variations of such worksheets. An FMEA can be a qualitative analysis but may be put on a quantitative basis when mathematical failure rate models are combined with a statistical failure mode ratio database. It was one of the first highly structured, systematic techniques for failure analysis. It was developed by reliability engineers in the late 1950s to study problems that might arise from malfunctions in military systems. An FMEA is often the first step of a system reliability study. A few different types of FMEA analyses may be carried out, such as a Functional, Design, or Process FMEA. Sometimes the FMEA is extended to FMECA (failure mode, effects, and criticality analysis) to indicate that criticality analysis is performed too. FMEA is an inductive reasoning (forward logic) single point of failure analysis and is a core task in reliability engineering, safety engineering and quality engineering. A successful FMEA activity helps identify potential failure modes based on experience with similar products and processes — or based on common physics of failure logic. It is widely used in the development and manufacturing industries throughout various phases of the product life cycle. Effects analysis refers to studying the consequences of those failures on different system levels. Functional analyses are needed as an input to determine correct failure modes, at all system levels, both for functional FMEA or Piece-Part (hardware) FMEA. An FMEA is used to structure mitigation for risk reduction based on either failure (mode) effect severity reduction or based on lowering the probability of failure or both. The FMEA is in principle a full inductive (forward logic) analysis, however the failure probability can only be estimated or reduced by understanding the failure mechanism. Hence, FMEA may include information on causes of failure (deductive analysis) to reduce the possibility of occurrence by eliminating identified (root) causes. The analysis should always start by listing the functions that the design needs to fulfil. Functions are the starting point of a well done FMEA and using functions as baseline provides the best yield of an FMEA. After all, a design is only one possible solution to perform functions that need to be fulfilled. This way an FMEA can be done on concept designs as well as detail designs, on hardware as well as software, and no matter how complex the design (Fig. 7.1).

When performing FMEA, interfacing hardware (or software) is first considered to be operating within specification. After that it can be extended by consequently using one

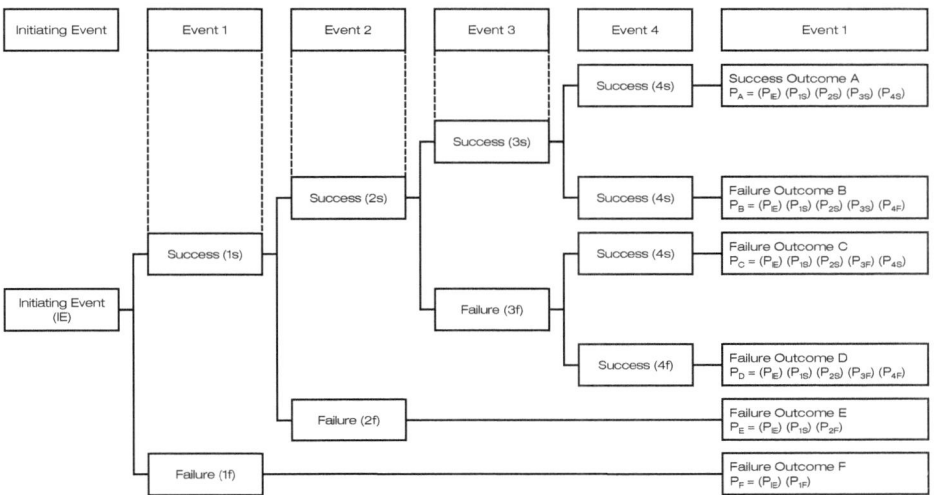

Fig. 7.1 Example of an event tree analysis

FMEA													

Process/Product Name _____   Prepared By _____
Responsible _____   FMEA Date (Orig.) _____   (Rev.) _____

Process Step/Input	Potential Failure Mode	Potential Failure Effects		Potential Causes		Current Controls			Action Recommended	Resp.	Actions Taken			
What is the process step, change or feature under investigation?	In what ways could the step, change or feature go wrong?	What is the impact on the customer if this failure is not prevented or corrected?		What causes the step, change or feature to go wrong? (how could it occur?)		What controls exist that either prevent or detect the failure?			What are the recommended actions for reducing the occurrence of the cause or improving detection?	Who is responsible for making sure the actions are completed?	What actions were completed (and when) with respect to the RPN?			

Fig. 7.2 Example FMEA worksheet

of the five possible failure modes of one function of the interfacing hardware as a cause of failure for the design element under review. This gives the opportunity to make the design robust for function failure elsewhere in the system. In addition, each part failure postulated is the only failure in the system (i.e., it is a single failure analysis). In addition to the failure mode and effects analysis' done on systems to evaluate the impact lower-level failures have on system operation, several other FMEAs are done. Special attention is paid to interfaces between systems and in fact at all functional interfaces. The purpose of these FMEAs is to assure that irreversible physical and or functional damage is not propagated across the interface because of failures in one of the interfacing units. These analyses are done to the piece part level for the circuits that directly interface with the other units. The FMEA can be accomplished without a criticality analysis (CA), but a CA requires that the FMEA has previously identified system level critical failures. When both steps are done, the total process is called an FMEA.

Some of the benefits derived from a properly implemented FMEA effort include the provision of a documented method for selecting a design with a high probability of successful operation and safety; a documented uniform method of assessing potential failure mechanisms, failure modes and their impact on system operation, resulting in a list of failure modes ranked according to the seriousness of their system impact and likelihood of occurrence; early identification of single failure points (SFPS) and system interface problems, which may be critical to mission success and or safety. They also provide a method of verifying that switching be- tween redundant elements is not jeopardised by postulated single failures; an effective method for evaluating the effect of proposed changes to the design and or operational procedures on mission success and safety; a basis for inflight troubleshooting procedures and for locating performance monitoring and fault-detection devices; and the creation of criteria for early planning of tests. From the above list, early

Fig. 7.3 Example FMEA
worksheet (Severity scale)

Severity Scale		
	Adapt as appropriate	
Effect	**Criteria: Severity of Effect**	**Ranking**
Hazardous - Without Warning	May expose client to loss, harm or major disruption - failure will occur **without** warning	10
Hazardous With Warning	May expose client to loss, harm or major disruption - failure will occur **with** warning	9
Very High	Major disruption of service involving client interaction, resulting in either associate re-work or inconvenience to client	8
High	Minor disruption of service involving client interaction and resulting in either associate re-work or inconvenience to clients	7
Moderate	Major disruption of service not involving client interaction and resulting in either associate re-work or inconvenience to clients	6
Low	Minor disruption of service not involving client interaction and resulting in either associate re-work or inconvenience to clients	5
Very Low	Minor disruption of service involving client interaction that does not result in either associate re-work or inconvenience to clients	4
Minor	Minor disruption of service not involving client interaction and does not result in either associate re-work or inconvenience to clients	3
Very Minor	No disruption of service noticed by the client in any capacity and does not result in either associate re-work or inconvenience to clients	2
None	No Effect	1

identifications of SFPS, input to the troubleshooting procedure and locating of performance monitoring/fault detection devices are probably the most important benefits of the FMEA. In addition, the FMEA procedures are straightforward and allow for the orderly evaluation of the system design.

7.7   FMEA Worksheet

For examples of FMEA worksheets, refer to:

- Fig. 7.2 Example FMEA worksheet
- Fig. 7.3 Example FMEA worksheet (Severity scale)
- Fig. 7.4 Example FMEA worksheet (Occurrence scale)
- Fig. 7.5 Example FMEA worksheet (Detection scale)

7.8   Probability (P)

It is necessary to look at the cause of a failure mode and the likelihood of occurrence. This can be done by analysis, calculations/FEM, looking at similar items or processes and the failure modes that have been document.

A failure cause is looked upon as a design weakness. All the potential causes for a failure mode should be identified and documented. This should be in technical terms. Some examples of causes are human errors in handling, manufacturing induced faults, fatigue, mission creep, abrasive wear, erroneous algorithms, excessive voltage, and improper operating conditions or use (depending on the used ground rules). A failure mode may be given

Fig. 7.4 Example FMEA
worksheet (Occurrence scale)

Probability of Failure	Time Period	Per Item Failure Rates	Ranking
Very High: Failure is almost inevitable	More than once per day	>= 1 in 2	10
	Once every 3-4 days	1 in 3	9
High: Generally associated with processes similar to previous processes that have often failed	Once every week	1 in 8	8
	Once every month	1 in 20	7
Moderate: Generally associated with processes similar to previous processes which have experienced occasional failures, but not in major proportions	Once every 3 months	1 in 80	6
	Once every 6 months	1 in 400	5
	Once a year	1 in 800	4
Low: Isolated failures associated with similar processes	Once every 1 - 3 years	1 in 1,500	3
Very Low: Only isolated failures associated with almost identical processes	Once every 3 - 6 years	1 in 3,000	2
Remote: Failure is unlikely. No failures associated with almost identical processes	Once Every 7+ Years	1 in 6000	1

Fig. 7.5 Example FMEA
worksheet (Detection scale)

Detection	Criteria: Likelihood the existence of a defect will be detected by process controls before next or subsequent process, -OR- before exposure to a client	Ranking
Almost Impossible	No known controls available to detect failure mode	10
Very Remote	Very remote likelihood current controls will detect failure mode	9
Remote	Remote likelihood current controls will detect failure mode	8
Very Low	Very low likelihood current controls will detect failure mode	7
Low	Low likelihood current controls will detect failure mode	6
Moderate	Moderate likelihood current controls will detect failure mode	5
Moderately High	Moderately high likelihood current controls will detect failure mode	4
High	High likelihood current controls will detect failure mode	3
Very High	Very high likelihood current controls will detect failure mode	2
Almost Certain	Current controls almost certain to detect the failure mode. Reliable detection controls are known with similar processes	1

a Probability Ranking with a defined number of levels. For a piece part FMEA, quantitative probability may be calculated from the results of a reliability prediction analysis and the failure mode ratios from a failure mode distribution catalogue, for example, RAC FMD-97.[6] This method allows a quantitative FTA to use the FMEA results to verify that undesired events meet acceptable levels of risk.

[6] https://armypubs.army.mil/epubs/DR_pubs/DR_a/pdf/web/tm5_698_4.pdf.

7.9   Severity (S)

Determine the Severity for the worst-case scenario adverse end effect (state). It is convenient to write these effects down in terms of what the user might see or experience in terms of functional failures. Examples of these end effects are:

- Full loss of function 'x'
- Degraded performance
- Functions in reversed mode
- Too late functioning, and
- Erratic functioning, etc.

Each end effect is given a Severity number (S) from, say, 'I' (no effect) to 'V' (catastrophic), based on cost and or loss of life or quality of life. These numbers prioritise the failure modes (together with probability and detectability).

7.10   Dormancy or Latency Period

The average time that a failure mode may be undetected may be entered if known. For example:

- Seconds, auto detected by maintenance computer
- 8 h, detected by turn-around inspection
- 2 months, detected by scheduled maintenance block 'x'
- 2 years, detected by overhaul task 'x'.

7.11   Indication

If the undetected failure allows the system to remain in a safe/working state, a second failure situation should be explored to determine whether an indication will be evident to all operators and what corrective action they may or should take. Indications to the operator should be described as follows:

- *Normal*. An indication that is evident to an operator when the system or equipment is operating normally
- *Abnormal*. An indication that is evident to an operator when the system has malfunctioned or failed
- *Incorrect*. An erroneous indication to an operator due to the malfunction or failure of an indicator (i.e., instruments, sensing devices, visual or audible warning devices, etc.).

7.12   Risk level (P × S) and (D)

Risk is the combination of end effect probability and severity where probability and severity include the effect on non-detectability (dormancy time). This may influence the end effect probability of failure or the worst-case effect severity. The exact calculation may not be easy in all cases, such as those where multiple scenarios (with multiple events) are possible, and detectability/dormancy plays a crucial role (as for redundant systems). In that case FTA and or ETA may be needed to determine the exact probability and risk levels. Preliminary risk levels can be selected based on a risk matrix based on MIL-STD-882. The higher the risk level, the more justification and mitigation is needed to provide evidence and lower the risk to an acceptable level. High risks should be indicated to higher level management, who are responsible for final decision-making. The FMEA should be updated whenever a new cycle begins (new product/process); changes are made to the operating conditions; a change is made in the design; new regulations are instituted; or customer feedback indicates a problem. The FMEA is useful for the development of system requirements that minimise the likelihood of failures; the development of designs and test systems to ensure that the failures have been eliminated or the risk is reduced to acceptable level; the development and evaluation of diagnostic systems; and helping with design choices (trade-off analysis). The utility of the FMEA should not be underestimated as the FMEA has many advantages, such as being a catalyst for teamwork and idea exchange between functions; collecting information to reduce future failures, capture engineering knowledge; providing early identification and elimination of potential failure modes; emphasising problem prevention; fulfilling legal requirements (product liability); improving company image and competitiveness; improving production yield; improving the quality, reliability, and safety of systems and processes; increasing user satisfaction; maximising profit by reducing impact on company profit margins; minimising late changes and associated costs; reducing system development time and costs; and reducing the potential for the same kind of failure in the future.

With that said, it is equally important to recognise the limitations of the FMEA. While FMEA identifies important hazards in a system, its results may not be comprehensive, and the approach has limitations. In the healthcare context, FMEA and other risk assessment methods, including SWIFT (Structured What If Techniques) and retrospective approaches, have been found to have limited validity when used in isolation. Challenges around scoping and organisational boundaries appear to be a major factor in this lack of validity. If used as a top-down tool, FMEA may only identify major failure modes in a system. FTA is better suited for "top-down" analysis. When used as a "bottom-up" tool FMEA can augment or complement FTA and identify many more causes and failure modes resulting in top-level symptoms. It is not able to discover complex failure modes involving multiple failures within a subsystem, or to report expected failure intervals of failure modes up to the upper-level subsystem or system. Additionally, the multiplication of the severity, occurrence and detection rankings may result in rank reversals, where a less serious failure mode receives a higher RPN than a more serious failure mode. The reason for this is that the rankings are ordinal scale numbers, and multiplication is not defined for ordinal numbers. The ordinal rankings only say that one ranking is better or worse than another, but not by how much. For instance, a ranking of "2" may not be twice as severe as a ranking of "1", or an "8" may not be twice as severe as a "4", but multiplication treats them as though they are. Various solutions to these problems have been proposed, for example, the use of fuzzy logic as an alternative to the classic RPN model. In the new AIAG/VDA FMEA handbook[7] (2022) the RPN approach was replaced by AP (action priority). In summary, the FMEA worksheet is hard to produce, hard to understand and read, as well as hard to maintain. The use of neural network techniques to cluster and visualise failure modes was suggested starting from 2010. An alternative approach is to combine the traditional FMEA table with a set of bow tie diagrams. The diagrams provide a visualisation of the chains of cause and effect, while the FMEA table provides the detailed information about specific events.

7.13   Bow Tie Analysis

Bow tie analysis or bow tie diagrams are a form of diagram which is used to model and visualise risk management and preparedness. The diagram visualises an event with its perceived threats, consequences, damage mitigation measures and preventive measures. Bow tie diagrams have been successful in assisting various industrial processes including in engineering, oil and gas, aviation, industrials, and finance. Bow tie analysis is believed to have originated from Imperial Chemical Industries (ICI) in the 1970s, though Royal Dutch Shell is the first major company to successfully integrate bow tie analysis into their business practices. Bow tie analysis is usually found in one of the last phases of the risk management process, after the assessment of risk, where specific events are focused and placed in the middle of the diagram. Threats and preventive measures are placed in the left part of the diagram, as part of the preventive controls for potential causes. Consequences and damage mitigation measures are placed on the right side, as part of the potential impact and corresponding plan for recovery preparedness. Bow tie analysis is sometimes simplified to exclude any barriers, i.e., to only illustrate a left and right side without preventive measures and damage mitigation measures.

In this chapter, we have explored the theory and application of different risk analysis methodologies including systems hazard analysis, fault tree analysis, failure modes and effects analysis and bow tie analysis (Fig. 7.6).

---

[7] https://www.aiag.org/store/publications/details?ProductCode=FMEAAV-1.

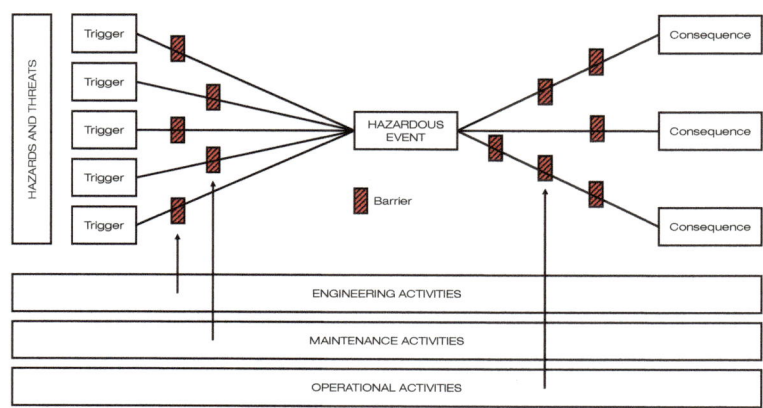

Fig. 7.6 Example of a typical bow tie analysis diagram

Risk Management and Hazard Reporting

8

The objective of this chapter is to describe the processes, methods and techniques used to manage risks once they have been identified; and, to ensure that risks are always reduced to acceptable levels throughout the lifecycle of a system. By risk management, we refer to the process by which risks, and hazards, are tracked appropriately so that mitigations can be applied to reduce the risk associated with the hazards to an acceptable level. In the UK this generally means reducing risks in accordance with the ALARP principle. This chapter examines some of the ways in which risks, and their hazards, are managed throughout the safety lifecycle, so that appropriate risk analysis and assessment techniques may be applied at the correct time, and that all risks and hazards are tracked through the system lifecycle.

8.1 Typical Hazard Lifecycle

A hazard may exist continuously as the result of a particular system operation or may be created as the result of a sequence of events. A further sequence of events is usually required for a hazard to lead to an accident. For instance, when cooking with gas the hazard 'exposed flame' is always present, but some further event, such as the spillage of cooking oil is required before the accident (i.e., the fire) occurs. A combination of accident severity and probability of occurrence leads to the assignment of a level of risk to the hazard/accident combination. Once identified, this risk level needs to be recorded, and any actions taken to mitigate (reduce) the risk also recorded. The normal vehicle for recording this information is the hazard log (covered in the next chapter). There are two main categories into which risks fall after initial risk allocation has taken place:

© The Author(s), under exclusive license to Springer Nature Switzerland AG 2025   71
A. A. Olsen, *Hazard and Risk Analysis for Organisational Safety Management*, Synthesis Lectures on Ocean Systems Engineering, https://doi.org/10.1007/978-3-031-73458-8_8

(1) Risks which are intolerable and therefore action must be taken, and

(2) Risks which may be considered tolerable; a judgement is required as to whether further risk reduction can sensibly be achieved.

This classification and any subsequent action taken is generally recorded in the project hazard log, in association with the appropriate hazard. In any real-world system, there will be residual risks which will need monitoring through the system lifecycle, and changes to the system will require review of these residual risk levels in the form of impact analysis and further hazard analysis and or risk assessment as required. This point cannot be stressed enough, as frequently it is the interaction of system elements each of which has a low level of residual risk associated with it, which may lead to an accident. Hazard analysis and risk assessment reports and the activities which underpin them have been described in earlier chapters (e.g., preliminary hazard analysis, risk assessment, etc.) The means by which the results of these analyses are recorded and tracked is the hazard log. This document acts as a repository of information relating to all hazards identified in relation to the system, the accidents associated with these hazards, the ways the hazards and accidents may arise, and the current level of outstanding risk associated with each accident occurrence. The hazard log remains a live document throughout the lifetime of the system to which it applies. Typically, once a system has been installed and is operational, all identified risks will have been mitigated to an acceptable level, but it is possible that new hazards will be revealed during operation or that system changes will lead to the introduction of new means by which risks may arise. The occurrence and resolution of all such issues will need to be recorded in the hazard log. Another issue that needs to be emphasised is that of transfer of responsibility for controlling risks. Once a system has been designed and handed over to operators, designers are in general no longer the most appropriate people to monitor risk. This task is, not surprisingly, now likely to be done by the operator. Other parties may also become involved. Therefore, the hazard log acts as a reference point for all these parties, identifying where residual risk remains, and those areas where significant risks have been mitigated.

8.2   Why Risk Management?

8.2.1   Case Study: The Nimrod XV230 Incident

On 2 September 2006, an RAF Hawker Siddeley Nimrod suffered an inflight fire and subsequently crashed in the Kandahar region of Afghanistan, killing all 14 crew members onboard. The crash, which occurred during a reconnaissance flight, was the biggest single loss of life suffered by the British military since the Falklands War (1982). The aircraft involved in the accident was XV230, the first of 38 Nimrod maritime reconnaissance/strike aircraft to enter operational service with the RAF on 2 October 1969. On 2

September 2006 the aircraft is believed to have suffered a fuel leak or overflow during mid-air refuelling while it was monitoring a NATO offensive against Taliban insurgents west of Kandahar. The investigation found that fuel most probably travelled from a fuel tank blow-off valve on the starboard side of the lower-forward fuselage into an aft bay located near the root of the starboard wing. This section of the wing contained hot air ducting pipes. It is thought the leaked fuel saturated the compressed insulation contained within the pipe shrouding, holding the fuel firm against a hot air pipe until it reached its auto-ignition temperature and caught fire. The fire was first noted when smoke accumulated in the bomb-bay, leading the pilot to report a fire in the aircraft's bomb-bay. The pilot tried to reach Kandahar Airfield, taking the aircraft down from 23,000 feet to 3,000 feet (7,010 to 910 m) in 90 s. A US Marine Corps AV8-B Harrier aircraft followed the Nimrod as it reduced its altitude before witnessing the starboard wing explode. This was followed a few seconds later by the rest of the aircraft. The crash site was about 25 miles (40 kms) west-north-west of Kandahar Airfield. Twelve RAF personnel, a Royal Marine and a British Army soldier onboard the Nimrod MR2 XV230 were killed.

The official board of inquiry report was released in December 2007. On 23 May 2008 the assistant deputy coroner for Oxfordshire, Andrew Walker, handed down a narrative ruling stating that the aircraft had.

"… never been airworthy from the first time it was released to the service [RAF] nearly 40 years ago … It seems to me that this is a case where I would be failing in my duty if I didn't report action to the relevant authority that would prevent future fatalities … I have given the matter considerable thought and I see no alternative but to report to the Secretary of State that the Nimrod fleet should not fly until the ALARP [as low as reasonably practicable] standards are met".

During the investigation into the Nimrod incident, it was found that concerns were raised previously when on 5 November 2007, Nimrod XV235 was reported to have suffered a similar fuel leak. In this instance the aircraft landed safely. The MOD suspended all inflight refuelling of the Nimrod fleet. In March 2009, following continued questions about the safety of the Nimrod fleet and despite constant protestations that the aircraft were airworthy, the MOD grounded the Nimrod fleet for "vital safety modification[s]". This entailed the wholesale replacement of the engine bay hot air ducts and fuel seals. Even so, concerns about the safety of the Nimrod fleet continued long after the loss of XV230. In April 2009 it was reported that the British Secretary of Defence had "glossed over Nimrod safety fears". The Independent newspaper claimed that a report into the safety of Britain's ageing fleet of Nimrod aircraft, which an anonymous British defence minister claimed did not reveal "any significant airworthiness issues", exposed in excess 1,500 faults—26 of which threatened the aircraft's safety. On 4 December 2007 the report of the findings by the official board of inquiry into the loss of XV230 was published. The board of inquiry believed that the No.7 tank dry bay was the most likely location for the seat of the fire. The probable cause being escaped fuel pooling around a Supplementary

Conditioning Pack (SCP) air pipe at some 400°C (752°F) "… after entering a gap between two types of insulation". The report listed four separate factors which contributed to the incident. These were:

(1) The age of the aircraft
(2) The RAF's maintenance policy
(3) A lack of a fire detection and suppression systems, and
(4) A failure to identify and recognise the full implications of successive changes to the fuel system and associated procedures.

In 2009 the long-awaited independent report into the Nimrod XV230 incident was published. The report, chaired by the respected judge Charles Haddon-Cave QC, highlighted several systemic and cultural failures within the MOD, BAE Systems, and QinetiQ. Tellingly, the Haddon-Cave Report asked two leading questions:

(1) Could the accident have been predicted and avoided?
(2) How could a history of failures have been recorded?

This mirrors the opinion given by the Oxfordshire coroner, Andrew Walker, who stated in his coroner report that the entire Nimrod fleet had "never been airworthy from the first time it was released to service" and urged that it should be grounded. Hadden-Cave added further "this cavalier approach to safety must come to an end. There were failures… [in monitoring the aircraft's safety]… that should, if the information had been correctly recorded and acted upon, have led to the discovery of this design flaw within the Nimrod fleet". The main outcome of the Hadden-Cave report was the demand that all past, current and future hazards are recorded in some form of official log (Fig. 8.1).

8.3   Hazard Log

The hazard log, as defined by Def Stan 00–56, is "the continually updated record of the hazards, accident sequences and accidents associated with a system. It includes information documenting risk management for each hazard and accident". The hazard log contains the traceable record of the hazard management process for the system or process in question, and subsequently:

• Ensures the project safety programme uses a consistent set of safety information
• Facilitates oversight of the project by a project safety committee and other stakeholders of the status of the safety activities
• Supports the effective management of possible hazards and accidents so that the associated risks are brought to and maintained at a tolerable level, and

Fig. 8.1 RAF Nimrod MR2 on patrol, North Pole. MOD © Crown Copyright

- Provides traceability of the safety decisions made.

These hazards, accident sequences and accidents are those which could potentially happen, and not only the ones which have already been experienced. The term hazard log is considered by some to be misleading as the information stored within the hazard log relates to the entire safety programme including accidents, controls, risk evaluation and ALARP justification, as well as data on hazards. It is important that any open or outstanding issues in the hazard log are regularly reviewed by the project safety committee to ensure that actions are completed, and any unacceptable risks are resolved accordingly. To be effective, the hazard Log should be treated as a live document and as such must be updated throughout the system lifecycle. The hazard log should be established at the initial stages of a project and remain current throughout the system lifecycle from inception to end of life decommissioning. At this point the hazard log can provide a record of all safety assessment information and evidence associated with a programme; a source of documentation relating to all safety risk evaluations conducted throughout the system life cycle; an auditable tracking mechanism for the system, showing what decisions were taken, when and why, and a cross reference to all other safety analysis and documentation for the system lifecycle. The hazard log should describe the system to which it relates, and record its scope of use, together with the safety requirements. When hazards are identified, the hazard log will show how these hazards were evaluated and the resulting residual risk assessed and will either recommend further action to mitigate the hazards, or formally document the acceptance of these hazards and the ALARP justification. In doing so the hazard log provides a structured way of storing and referencing safety risk evaluations and other information relating to an equipment or system and should be coordinated and controlled whilst maintaining an auditable record of that information. In essence, it is the principal means of tracking the status of identified hazards, decisions made, and actions undertaken to reduce risk and should be used to facilitate oversight by key stakeholders.

Given the hazard log is a tracking system for hazards, their closures, and residual risk, it should be maintained throughout the system lifecycle as a "live" document. As changes are integrated into the system, the hazard log should be updated to incorporate new or amended status hazards and the associated residual risk to reflect the current design standard. The hazard log should capture the inputs to and outputs from hazard analysis and risk evaluation activities. ALARP justification arguments and conclusions should be recorded when mitigation actions are completed. Despite there being no universally accepted structure or approach to the layout of the hazard log, a basic hazard log is expected to contain as a minimum the following sections and content:

- Part 1: Introduction
- Part 2: Accident data
- Part 3: Hazard data
- Part 4: Statement of risk classification, and

- Part 5: Journal.

8.3.1 Part 1: Introduction

Part 1: Introduction should describe the purpose of the hazard log and indicate the environment and safety criteria to which the system safety characteristics relate. The following details, appropriate to the programme phase, should be contained in this part:

- The purpose and structure of the hazard log. This should be of sufficient detail to ensure that all project personnel understand the aim and purpose of the hazard log. The procedure for managing the hazard log should also be included
- A description of the system and its scope of use. This should include reference to a unique system identifier
- Reference to the system safety requirements
- The accident severity categories, probability categories, equivalent numerical probabilities, and accident risk classification scheme for the system
- The design rules and techniques for each SIL
- The apportionment of the random and systematic (SIL) elements of the hazard probability targets between all the functions of the system.

The description and scope of use of the system will be stated to indicate the environment to which the system safety characteristics relate. This information should be entered in part 1 of the hazard log.

8.3.2 Part 2: Accident Data

Part 2: Accident data should give sufficient information to identify the accident sequence linking each accident and the hazards which may cause it. It should typically include:

- A unique reference
- A brief description of the accident
- The accident severity category and probability targets appropriate to Risk Classes B and C
- A cross reference to the full description and analysis of the accident sequence in the safety programme reports. This information should be used to justify the subsequent setting of the hazard probability targets, and
- A list of the hazards and associated accident sequences that can cause the accident.

8.3.3   Part 3: Hazard Data

Part 3: Hazard data should provide sufficient information to identify the risk reduction process applicable to a particular hazard. A summary of all hazards and their status, including any outstanding corrective actions, should be contained within part 3 to provide an overview of the current situation. Part 3 should contain the following information for each hazard:

- A unique reference number
- A brief description of the hazard which should comprise the functions or components and their states that represent the hazard. Reference should also be made to the design documentation which describes the functions or components
- The related accident severity category, and the random and systematic elements of the hazard probability targets appropriate to Risk Classes B and C
- The predicted probability for the random element of the hazard
- A statement as to whether the hazard requires further action to reduce the risk from the system to a tolerable level
- A discussion of any possible means by which the risk could be reduced to a tolerable level, and notes on the re-evaluation of the accident sequence following such action
- A brief description of the action to reduce risk, together with either a reference to the design documentation that has changed because of the action, or the justification for taking no action
- A cross reference to the full description and analysis of the hazard in the hazard analysis reports.

8.3.4   Part 4: Statement of Risk Classification

A statement of risk classification should be included in Part 4: Statement of risk classification to provide a brief statement of the current system risk class. This should contain sufficient information to enable it to be a standalone statement, and it should contain the hazard log reference to enable traceability to its supporting documentation.

8.3.5   Part 5: Journal

Part 5: Journal should consist of a regularly updated journal which provides a historical record of the compilation of the hazard log. The journal should contain the following information:

- The date the hazard log was started

- All entries made in the hazard log, including any accident or hazard reference numbers
- Reference to the safety programme plan.

The process for a hazard log requires several initial steps to be undertaken prior to populating the hazard log. This is to ensure that there is a suitable infrastructure in place before hazard information is stored: (1) a method through which the hazard log is to be implemented should be selected; this can either be in paper or electronic form. It is important at the outset to identify the appropriate tool(s)/administration method for maintaining the hazard log; (2) a hazard log owner should be appointed, with delegated responsibility for the maintenance, upkeep, and configuration control of the hazard log. If the hazard log is kept in an electronic format, all non-owners should be allowed read-only access. (3) the hazard log should be 'set up'. This includes activities such as the inclusion of the risk classification scheme that has been agreed, the determination of appropriate hazard categories, status definitions and general set up activities to ensure the hazard log will operate as required. The latter may be in the form of guidance for a paper-based system or checking the robustness of an electronic system; (4) once the system and its boundaries have been defined and the hazard identification process has begun, the hazard log should be established to keep a record of the hazards and proposed or implemented mitigation measures to ensure that the hazards are being appropriately controlled. The hazard log should be the configuration control mechanism for the safety assessment process, and hazards should not be deleted from the hazard log, but instead closed out and marked if no longer relevant; (5) a procedure should be defined for the management and control of the hazard log. The hazard log should be retained for the entire system lifecycle, and it should act as the primary source of the logical arguments, or Safety Case, for the deployment of the system into service; (6) the hazard log should be reviewed at regular intervals to ensure that hazards are being successfully managed and that the robustness of the established safety arguments in the Safety Case are not being compromised. Generally, the hazard log should be updated whenever:

- A relevant hazard or potential accident is identified, either through formal analysis or because of a change to the design, procedure or operating environment
- A relevant incident occurs, perhaps during testing or demonstration
- Further information relating to existing hazards, incidents or accidents comes to attention; or safety documentation is created or re-issued.

To provide project awareness of hazard and accident data, the hazard log should be accessible to all appropriate project personnel. The hazard log must be available for inspection by the safety auditor, the safety assessor and representatives of any relevant safety authorities or defence regulators (as appropriate). As the hazard log is a repository for managing identified hazards, it is possible for hazard identification to begin prior to the implementation of the hazard log. Once the initial steps have been undertaken, the process of information entry can be started. The generic flow of the process is shown in the following steps:

- **Hazard Identification.** This is initially taken from the PHA, and should then be augmented by subsequent risk management activities
- An **accident sequence** should be developed in associated with the identified hazards
- A formal **Risk Evaluation** of each accident sequence should occur
- **Mitigation identification.** The appropriate and agreed mitigation for each accident should be recorded
- **Mitigation/control owners established.** This should ensure that the mitigation or controls identified are put in place and the hazard is addressed
- **Cross checking** should take place to see if there are any other, previously identified hazards or accident sequences linked with this hazard
- **Resolution.** Status changes should be completed as required, formal closures recorded, including reference to evidence and ALARP justification recorded
- **Ongoing hazards** should be managed, and new hazards added as required
- A **Hazard Log Report** as determined by the project safety plan should be produced.

Where the project safety programme identifies hazards that are the responsibility of another project, then the information should be passed to the person with delegated authority for that area. The hazard log should record that this was done. Because hazards and accidents usually have a range of control measures of different types associated with them, there is no single hazard "owner" who is responsible for mitigating the associated risks, other than the overall delegated authority. When a control measure is agreed for implementation, it should be clearly assigned to an "owner". This might be the prime contractor for a design change, the training authority for a topic to be covered in maintainer training, or the user for a procedural control solution.

8.4   Closure and Removal of Entries

It is considered best practice for the hazard log to record each hazard as "open" and for ALARP arguments to be provisional until all mitigation actions are confirmed to be satisfactorily completed. An example is where the mitigation depends upon production of an operational procedure that may not be written for a considerable time after the hazard is first identified at an early stage of design or construction. However, equipment should not be declared operational (or used in any scenario where it may present a hazard, e.g., trials) with risks that have not been formally declared ALARP. In the Military context, this may exceptionally mean declaring ALARP with mitigation measures outstanding. In such circumstances, and only when strict guidelines have been followed, ALARP may be declared. Hazards should not be deleted from the hazard log but closed and marked as "out of scope", "not considered credible" or some other alternative, together with the justification. Where they are no longer considered relevant to the system, the log entry should be updated to reflect this. At the end of the project, the hazard log should remain as a historical record, which should be useful to refer to for similar applications in the future.

8.5   Record Keeping and Project Documentation

Adequate provision should be made for the security and backup of the hazard log and other safety records. The hazard log is a prime source of corporate knowledge and is the configuration control mechanism for the safety assessment process. As such it could be referred to in legal proceedings. Every effort should therefore be made by the project to ensure that records are accurate, attributable, up to date and complete. Clear cross-referencing to supporting documents is essential. Where relevant, the outputs from this procedure should feed into the System Requirements Document—for any specific safety requirements; the customer supplier agreement—to document agreements on safety information to be delivered by the delivery team; the through life management plan; and the safety elements of outline business case and full business case submissions. The relationship between hazards, accidents and their management through setting and meeting safety requirements could be included within the hazard log. However, if it is not sufficiently robust or well-structured, this may overload the hazard log and obscure the identification and clearance of hazards. Good management and oversight of the hazard log is an important part in demonstrating the robustness of evidence of safety and should be clearly documented and referenced. If hazards are not well defined when they are entered into the hazard log, then the rigour enforced by the need for a clear audit trail of changes made, may make it very difficult to maintain the hazard and accident records in the most useful structure. An appropriate structure should therefore be designed and agreed before data entry starts.

8.6   Procedure Completion

The hazard log ensures that a common set of information can be shared by all parties with a genuine need for access. A single hazard log should therefore be maintained that is accessible by all these parties. The hazard log may be run by a prime contractor, the system owner or a third party such as a safety assessment contractor. Indeed, the hazard log may pass from one authority to another at key stages in the programme. For example, the prime contractor is likely to have greatest need of the hazard log during system development, but the system owner may be a more appropriate controller when the system is in service. The hazard log should remain under the control of a hazard

log owner, who is responsible to the system owner's safety management. The hazard
log owner should have full access to the hazard log allowing them to add, edit or close
out hazards. All other personnel requiring access to the hazard log should be assigned
read only access. This allows for visibility of hazards to all, but the strict control and
administration of hazards is limited to the hazard log owner. The hazard log should be
established at the earliest stage of the programme and be maintained thereafter as a 'live'
document or database to reflect the current design standard. Review of the hazard log is
essential at regular intervals to ensure that hazards are being successfully managed and
that the robustness of the safety arguments in the safety case can be established.

8.7   Hazard Log Inputs and Outputs

The procedure for hazard log requires inputs from the safety initiation; preliminary hazard identification and analysis; hazard identification and analysis; risk estimation; risk and
ALARP evaluation; risk reduction; risk acceptance; and safety requirements and contracts.
The hazard log is a database which references all the major items of safety documentation relating to a project. This may include the safety criteria report; safety requirements;
hazard identification reports; hazard analysis reports; risk analysis and assessment reports;
safety audit and inspection reports; and or safety case reports. The hazard log should store
information on hazards, accidents and accident sequences which might be associated with
the system. Hence, it records the results of all the Risk Management procedures. Where
the hazard log has adequate capacity and resources permit, the following supporting documentation should also be either directly embedded or cross-referenced by hypertext link
where the hazard log is kept in an electronic format: material/system survey reports;
design defect reports, concessions, and production permits; system/equipment breakdown
and failure reports; reports of technical design/material state reviews; reports of quality,
reliability, and safety audits; and accident and incident reports, during construction, maintenance, or in-service operation. The hazard log should be a continuously evolving record
(database or document) which should stay with the system throughout its life cycle. A
hazard log report is a snapshot of the hazard log status at any given time. Hazard log
reports are produced for the purpose of review—for example, by the project safety committee or the independent safety auditor or communication of the status of the safety
programme. Hazard log reports should be capable of showing the linkages between hazards, accidents and controls (i.e., which hazards could lead to which potential accidents,
possibly with many-to-many relationships, and which controls relate to which hazards and
accidents). They should also differentiate between controls which are already in place and
those which are being considered or planned.

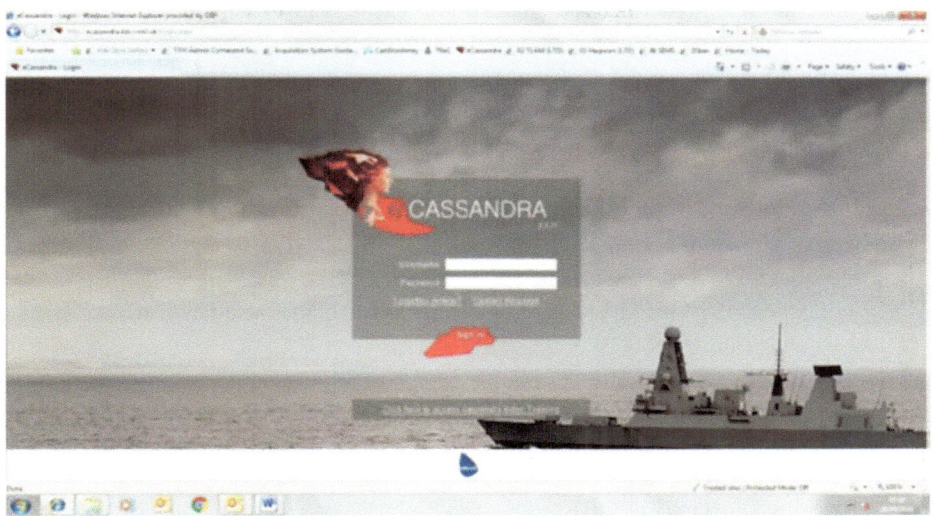

Fig. 8.2   eCassandra (MOD use only). MOD © Crown Copyright

8.8   Hazard Log Software

There is a multitude of hazard log software platforms available on the market today. Within the Defence realm, the mandated corporate hazard log tool is the Cassandra Database System and its online equivalent, eCassandra. Project leaders should consider tailoring their system to meet specific project needs. Moreover, individual projects or peer groups may develop additional database or hazard management programmes for their use, which are again tailored to suit individual project needs, provided that the solution satisfies the objectives of hazard log recording. Whichever hazard log tool is adopted, it should be under strict configuration control to maintain a robust audit trail (Fig. 8.2).

8.9   Limitations of the Hazard Log

The hazard log has faced increasing criticism for several reasons. These include:

- **The hazard log being too expansive.** It is not uncommon for complex systems, processes and products to have hazard logs that run to 000s of entries ranging from the high level (e.g., a car out of control) to the minute (e.g., insufficient oil). Too much inconsequential detail can hide potentially major failures in the design and development of the system, leading to future accidents. To avoid this issue, it is important to recognise the distinction between hazard and cause, and to collate related causes of hazards

- **The hazard log lacking detail.** Conversely, a hazard log that is too light on detail can lead to misinterpretations and even hazards being missed altogether. A light hazard log is often symptomatic of poor oversight, lack of analysis, and a lack of input information such as justifications
- **Failure to close out hazards.** When a hazard has been reduced to a tolerable level, but the hazard log remains unchanged, this can lead to wasted effort attempting to resolve a hazard that no longer exists. This can be at the expense of equally important hazards or a collection of lower-level hazards which, when combined, have the potential to cause a major catastrophe
- **Poor management and lack of control.** When too many people are responsible for maintaining the hazard log, the potential for mismanagement increases exponentially. To avoid this, it is important that a single point of contact is made responsible for overseeing the hazard log. This may be a delegated responsibility or a role-based responsibility
- **Lack of traceability.** The hazard log is a legal document that may—and often is— used by the law courts as evidence. This means the hazard log must be maintained as an authoritative source of information including the justification of acceptance of risk. A hazard log that is poorly maintained lacks this authority, breaking the chain of traceability.

8.10   DRACAS

The DRACAS is a closed loop data system for reporting and analysis, used to record information about incidents and corrective actions that have been implemented. The system is intended to provide confidence in the accuracy of the claimed theoretical analysis and correct operation of the safety features. DRACAS is a necessary part of any good management system that aims for continual improvement by tracking and resolving reported problems. The safety aspect of DRACAS relates to the elimination of failures and deficiencies which impact on safety and DRACAS can be a major con132.

8.11   Hazard and Risk Analysis

Tributor to achieving in-service safety. Over time, the scope of the closed-loop reporting has expanded from Failures to Faults (both abbreviated to FRACAS) to Defects and finally Data (DRACAS). This is in recognition that learning opportunities to improve the system, including its safety, encompass near misses, human errors, and other incidents as well as equipment malfunctions. The DRACAS should provide traceability of incident management from initial discovery to resolution, or until associated risks have been reduced to a tolerable level. DRACAS should cover all non-conformances and failures relating to design, manufacturing, use, maintenance, and test processes that occur during any project activity. Operational and usage data, together with operating conditions should also be recorded to allow event frequency rates to be estimated. DRACAS provides for reporting of suspected failures and non-conformances as well as observed failures, failure indications and non-conformances. Examples of inputs are incidents resulting from component failures, human error, etc., whether they cause harm. To provide effective management of non-conformances and failures, there should be provision in the DRACAS to ensure that effective action is taken promptly. The DRACAS should be audited to review all active reports, the failure and incident analyses, and corrective action deadlines. DRACAS should be established and maintained throughout the design, development, production and in-service phases of a project life cycle. The safety programme plan should document the requirement for DRACAS, either as a stand-alone activity or as part a Repair and Maintenance programme. All incidents and corrective actions recorded in DRACAS should be reviewed by a competent project safety engineer. If a new hazard is identified, or there is an amendment to the risk classification of an accident, the output from DRACAS must be fed into the hazard log. Information from and the use of DRACAS, along with a comprehensive test programme can provide evidence for safety verification. DRACAS provides several advantages, such as DRACAS encompasses all aspects of development tests, trials, and in-service usage; the use of DRACAS from early in a project aids design, identify corrective actions and help evaluate test results to improve safety performance, and the use of DRACAS aids elimination of failures and deficiencies, therefore supporting achievement of safety growth. Like all things, DRACAS also has several limitations, including data collection is only as good as the information fed back. Too complicated or detailed a request for information may result in DRACAS being perceived as being too onerous by the "reporter" and hence the system may not be fully utilised. This could give rise to a false sense of security, and data information and analysis should be managed by personnel who have the appropriate level of knowledge and experience. The example of a DRACAS process shows how an incident is managed, from the initial reporting of the incident to inclusion of the relevant information in the DRACAS database. The development and structure of the database is important in the successful application of a DRACAS. The need for tracking of incidents, comparison, providing historical evidence etc., means that a bespoke database may be preferable for certain usages.

In this chapter, we have explored the role and function of the hazard log, the information that should be uploaded and managed in the hazard log, and the importance of maintaining the hazard log as an authoritative record of hazards and risks. We have also briefly looked at DRACAS/FRACAS, and how these systems aid in the management of hazard recording and reporting. In the next and final chapter, we will turn our attention to safety arguments and safety cases.

Safety Arguments and Safety Cases

Following a series of fatal accidents (*Piper Alpha* in 1986 and Clapham Junction in 1988[1]) the UK put in place new legislation concerning safety. This legislation, which is based on the precept that it is an operator's responsibility to demonstrate that their activities are safe enough rather than the approach of complying with a detailed prescriptive set of requirements, represented a fundamental change in industrial safety practices. The acceptable safety of a system can be argued based on the claim that it has been developed to the current 'best practice' in safety engineering. Best practice is typically seen as embodied in the most recent safety standards and guidelines. One of the most significant safety standards for electronic systems and software is IEC 61,508, which specifies requirements for the functional safety of safety-related electrical / electronic and programmable electronic systems. The standard is generic, in that it applies to a range of industries and different safety applications and is applied in many countries. In this chapter, we will briefly

---

[1] The Clapham Junction railway crash occurred on the morning of 12 December 1988, when a crowded British Rail passenger train crashed into the rear of another train that had stopped at a signal just south of Clapham Junction railway station in London, England, and subsequently sideswiped an empty train travelling in the opposite direction. A total of 35 people died in the collision, while 484 were injured. The collision was the result of a signal failure caused by a wiring fault. New wiring had been installed, but the old wiring had been left in place and not adequately secured. An independent inquiry chaired by Anthony Hidden QC found that the signalling technician responsible had not been told that his working practices were wrong, and his work had not been inspected by an independent person. He had also performed the work during his 13th consecutive seven-day working week. Hidden was critical of the health and safety culture within British Rail at the time, and his recommendations included ensuring that work was independently inspected and that a senior project manager be made responsible for all aspects of any major, safety–critical project such as re-signalling work. British Rail was fined GBP 250,000 for violations of health and safety law in connection with the accident.

© The Author(s), under exclusive license to Springer Nature Switzerland AG 2025   87
A. A. Olsen, *Hazard and Risk Analysis for Organisational Safety Management*, Synthesis Lectures on Ocean Systems Engineering, https://doi.org/10.1007/978-3-031-73458-8_9

examine some of the key regulations and events in British history that have led to the development of the safe argument and safety case.

9.1   Robens Report

In 1970 Alfred Robens, Lord Robens of Woldingham, was set the task of scrutinising and overhauling a mountain of outmoded safety laws and regulations, which had been accumulating on the English statute books since around 1860. The Robens Report of July 1972 recommended the repeal of the bulk of this regulation and the advocation of its replacement with a single enabling Act of Parliament. In response to Lord Roben's recommendation, the British Government put in place the HASAWA in 1974. The underlining philosophy underpinning the HASAWA was that improved safety standards could be achieved through mechanisms such as self-regulation within a framework of minimal statutory requirements. This is because Robens believed "there are severe practical limits on the extent to which progressively better standards of safety and health at work can be brought about through negative regulation by external agencies. We need a more effectively self-regulating system. This calls for the acceptance and exercise of appropriate responsibility at all levels within industry and commerce. It calls for better systems of safety organisation, for more management initiative and for more involvement of work people themselves." Hence, the HASAWA sets the basic minimum requirements for health and safety at work, leaving the implementation of health and safety to individual employers and employees. It is worth noting that although the HASAWA remains the key piece of legislation relating to health and safety in the workplace, it has since been augmented by additional statutes and regulations relating to specific operational and health and safety related matters (often referred to as the "six pack" regulations:

- Management of Health and Safety at Work Regulations 1999
- Provision and Use of Work Equipment Regulations (PUWER) 1998
- Manual Handling Operations Regulations 1992
- Workplace (Health, Safety and Welfare) Regulations 1992
- Personal Protective Equipment at Work Regulations 1992
- Health and Safety (Display Screen Equipment (DSE) Regulations 1992.
- Several other key regulations and statutory instruments which apply include:
- Acetylene Safety (England and Wales and Scotland) Regulations 2014 (SI 2014/1639)
- Chemicals (Hazard Information and Packaging for Supply) Regulations 2002
- Confined Spaces Regulations 1997 (SI 1997/1713)
- Control of Major Accident Hazards Regulations 2015
- Control of Noise at Work Regulations 2005
- Control of Substances Hazardous to Health (COSHH) Regulations 2002
- Control of Vibration at Work Regulations 2005

- Dangerous Substances and Explosive Atmospheres Regulations 2002
- Dangerous Substances in Harbour Areas Regulations 1987
- Electricity at Work Regulations 1989
- Lifting Operations and Lifting Equipment (LOLER) Regulations 1998
- Pressure Equipment Regulations 1999
- Pressure Systems Safety Regulations 2000 (PSSR, SI 2000/128), replacing the earlier Pressure Systems and Transportable Gas Containers Regulations 1989
- Reporting of Injuries, Diseases and Dangerous Occurrences (RIDDOR) Regulations 2013
- Work at Height (WAH) Regulations 2005.

9.2   Flixborough Disaster, 1 June 1974

The Flixborough disaster was an explosion at a chemical plant close to the village of Flixborough, North Lincolnshire, England on Saturday, 1 June 1974. It killed 28 and seriously injured 36 of the 72 people on site at the time. The casualty figures could have been much higher if the explosion had occurred on a weekday when the main office area would have been occupied. The disaster involved (and may well have been caused by) a hasty equipment modification. Although virtually all the plant management personnel had chemical engineering qualifications, there was no onsite senior manager with mechanical engineering expertise. Mechanical engineering issues with the modification were overlooked by the managers who approved it, and the severity of potential consequences due to its failure were not accounted for. A plant modification occurred without a full assessment of the potential consequences. Only limited calculations were undertaken on the integrity of the bypass line. No calculations were undertaken for the dog-legged shaped line or for the bellows. No drawing of the proposed modification was produced. No pressure testing was carried out on the installed pipework modification. Those concerned with the design, construction and layout of the plant did not consider the potential for a major disaster happening instantaneously.

9.3   Health and Safety at Work Etc. Act 1974

Section 40 of the HASAWA places the burden of proof, in the event of a breach of the Act, on the offending party. This means the defendant (if in a court of law) must prove on the balance of probabilities that they did everything reasonably practicable to avoid the incident or accident. This is not an easy burden to discharge. The HASAWA is perhaps the most important development in industrial safety legislation of the twentieth century. It introduced a new approach to health and safety with an emphasis on self-regulation and employee participation. The HASAWA applies to all workplaces, irrespective of the

Fig. 9.1 Clapham Junction disaster (1988)

type of business. Prior to the HASAWA, the approach to industrial safety had been based on legislation that had grown in a piecemeal fashion over the preceding hundred or so years. Acts of Parliament were passed in response to hazards and work-related incidents as they arose. The HAWASA changed the direction of approach meaning employers (and employees) were now responsible for ensuring work practices are intrinsically safe to prevent incidents from occurring. The HASAWA has proven so successful in ensuring workplace safety that it has been adopted, in various forms, around the world (Figs. 9.1, 9.2, 9.3 and 9.4).

9.4   Corporate Manslaughter and Corporate Homicide Act 2007

The Corporate Manslaughter and Corporate Homicide Act 2007 entered into force on 6 April 2008. Before the enactment of the legislation, corporate manslaughter or corporate homicide could only be attributed to individual persons. This made it extremely difficult to prosecute large organisations and companies if an employee or member of the public were to sustain fatal injuries as a direct or indirect consequence of the organisation's actions. The aim of the act was to level the playing field between smaller businesses and

Fig. 9.2 Clapham Junction disaster (1988)

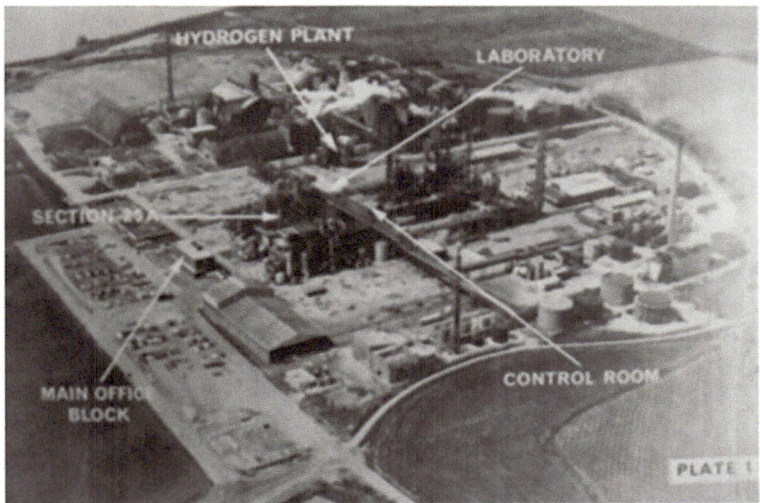

Fig. 9.3 Flixborough Plant before the explosion (official report, TS 84/37/1)

Fig. 9.4 Flixborough Plant after the explosion (official report, TS 84/37/1)

larger companies as with the former, it is not uncommon for the company director(s) to also be the company's owner, and often only employee. The new legislation made it easier to prosecute any corporate entity—irrespective of size or legal status—in the event of a fatality resulting from the gross negligence of the company's senior management. In effect, the new law made management personally and severally responsible for the actions of their organisation.

9.5   Permissioning Regimes

Both Robens (1972) and Cullen (1990) concluded that there was too much prescriptive regulation constraining British industry. They feared there was a danger that the existing regulatory regime could be complied with, but that this did not necessarily translate into a reduction in risk. The result was a move away from prescriptive regulation to a 'risk based' approach and the implementation of permissioning regimes. In essence, this placed the onus of responsibility on 'duty holders' to identify hazards and to assess risks. Having done so, the duty holder must demonstrate to the appropriate regulator that the organisation's activities are intrinsically safe. The main driver for moving away from pre-scription to a 'risk based' approach was the 1990 Cullen Report, which was published following the *Piper Alpha* tragedy in 1988. Cullen recommended that operators should be required—by regulation—to submit a safety case for each oil and gas installation to the Offshore Regulatory Authority. In doing so, the safety case should demonstrate that certain objectives had been met, including:

(1) The creation and implementation of safety management systems for the installation, which are adequately robust to ensure that the design and operation of the installation and its equipment is safe
(2) The identification of potential hazards of the installation and risks to personnel, and the imposition of appropriate controls
(3) The provision of procedures and protocols ensuring - in the event of a major emer-gency affecting the installation—that (a) a temporary safe refuge for personnel is provided on board the installation; and (b) enabling of safe and full evacuation measures for escape and rescue.

The safety case, as put forward as a recommendation of the 1990 Cullen Report, is a structured argument, supported by a body of evidence, which provides a compelling, comprehensive, and valid case that a system is safe for a given application in a given operating environment. Common themes amongst the standards and regulations concern-ing the creation and use of safety cases are the creation of evidentiary documentation; the implementation of valid and appropriate safety management systems; the identification of risks and hazards in accordance with the ALARP principle; and the management of hazards through a hazard log. Importantly, not all industries have safety case regulations. Those that do tend to operate within especially hazardous environments such as the off-shore and rail sectors. Alternatively, other sectors (such as the civil avionics industry) requires certification but do not a safety case.

9.6   Models for the Construction of a Safety Argument

The term 'safety argument' is used to encompass a justification that a particular piece of equipment, procedure, subsystem, system, etc. is acceptably safe. A safety argument is contained in a document that summarises the safety activities carried out and their results, during the development of a system, and presents a coherent and reasoned claim for the deployment of the system on the grounds of its accept- able safety. Safety arguments should refer to the detail of the documentation produced during the development process as background evidence to support the claim. Safety arguments need to be tailored to meet the specific objectives of the safety requirements and should highlight the key aspects of 'why the system is safe'. Safety arguments may encompass a justification for the use of anything from a fully operational plant down to a small component.

The term 'safety case', although having a generic meaning is also a special term with regulatory and legal connotations. In generic terms, a safety case is a term used by regulations to refer to a clear, defensible, objective, comprehensive and convincing argument, developed by the system operator that the system is acceptably safe. The safety case identifies the risks inherent in operating the system, demonstrates that the operating risks are fully understood, that they have been reduced to an acceptable level, and that they are properly managed. The analysis of risk may encompass the design, operation, inter-operation, change of use and decommissioning of the system. Within the UK the term 'safety case' is also used to represent statutory submissions made by operators to a regulator or body representing a regulator. The term is used for submissions in the offshore, rail, defence, and gas industries although submissions of similar form, but differently named, are made within the nuclear and chemical industries. 'Acceptably safe' is defined in terms of criteria laid down by a company, statutory body, regulator or even because of legal outcomes. Consequently, a safety argument can encompass a wide range of engineering systems and technology as well as procedural issues and can refer to widely differing safety criteria. Standards such as IEC 61,508 offer little advice, or even a structure, for collecting the evidence available from the development process and presenting a well-founded, well-justified, logical argument that the system has achieved a certain level of safety.

As a result, there are two dangers: the first is that the safety arguments presented for a particular system may not be in accordance with industry norms; there is a danger either of too little evidence in which case the organisation is at risk from negligence, or of too detailed evidence in which case the cost may be onerous; an additional danger is that over specification of safety has the effect that future activities must reach higher standards, hence escalating the cost base of the industry; and second, that industry is inefficient in the preparation of safety arguments; much of the development guidance requires the production of considerable information; without a clear view from the outset as to how the information will be used to justify the safety of the software, the volumes of information make it difficult 'to see the wood for the trees' and to present a cohesive safety argument. Therefore, for industry to provide safety arguments, which are acceptable to the regulators, further guidance is required on what evidence should be supplied with a safety argument. Such guidance is available in other standards, for example, Def Stan 00–56 and BS EN 50,126/9.[2]

---

[2] The standard EN 50,126 (in addition to EN 50,128 and EN 50,129) provides railway duty holders and railway suppliers with a process that enables the implementation of a consistent approach to the management of Reliability, Availability, Maintainability and Safety (RAMS).

9.7   Constructing the Safety Argument

The overall argument for the safety of a system must be based on the concept of the level of risk associated with the hazards of deployment of the system into its environment. Naturally enough, an overall system safety argument is related to the structure of the safety lifecycle. The main requirements of a safety argument include evidence to demonstrate that the system safety requirements are correct and complete; the system architecture is adequate; the safety requirements have been allocated correctly between sub-systems; and each subsystem is safe. Other essential elements of a safety argument include aspects such as safety management, competence, and independent assessment. These aspects are taken outside of the technical elements of our argument since they apply across all phases of the safety lifecycle and underpin all the safety argument. The activities that contribute towards demonstrating that the safety requirements are correct and complete comprise:

(1) Hazard identification
(2) Hazard analysis
(3) Risk assessment
(4) Safety requirements validation.

Note that the arguments for the completeness and correctness of the safety requirements include issues regarding their validation and will embody requirements for the operation, maintenance and potentially decommissioning of the system. There is somewhat of a 'chicken and egg' situation with the question of whether the system architecture can meet the safety requirements and the allocation of safety-requirements to subsystems. When developing a system, the overall question of the safety of the architecture cannot be answered before the safety requirements have been allocated. Similarly, safety-requirements should not be fixed until one is confident that the system architecture can provide a system that will satisfy the safety requirements. Therefore, the process to be adopted is iterative. IEC 61,508 Part 1 Clause 7.6 and the hardware architecture requirements in Part 2 are similarly linked.

In terms of overall performance of the safety system, its architecture is the most important of the technical issues to be addressed in the argument. If the system architecture is not adequate then, no matter the level of integrity of the individual subsystems, the overall safety requirements will not be met. Careful selection of the system architecture can significantly reduce the amount of effort required to develop individual subsystems and to demonstrate that they are safe. Regarding allocation of safety requirements, it is important that the interfaces between subsystems and with the wider environment are well-defined. Where there is a safety function that is implemented by a combination of functions in two subsystems, there may be issues regarding the independence of the subsystems. Experience has shown that, in the development of integrated systems, especially where there are different groups responsible for the supply of different subsystems, requirements can be wrongly assumed to have been met by another subsystem. A clear definition of shared interfaces between subsystems is essential, and for the purposes of a safety argument it is important that the responsibilities for implementation of safety functions have been assigned and that there is little possibility of a common cause failure which can disable two safety-related subsystems simultaneously. Common cause failures can be related to errors due to similar misinterpretation of the specification of a safety function which is shared across two sub-systems; similar implementations of safety functions that fail in the same way at the same point in time; shared inputs to (e.g., power supply or sensors), or outputs (e.g., signals and / or actuators) from, safety-related subsystems; and the physical placement of subsystems.

9.8   Drafting the Safety Case Report

The key deliverable of the safety case is the safety case report, which provides a summary of the safety arguments at any given time. There are three stages to the development of the safety case report. These are (a) the development of the safety strategy, which takes place during the requirements analysis phase of the project; (b) as an output of the systems hazard analysis and in preparation of safety validation; and (c) on completion of systems testing and in advance of the system certification and use. The creation of safety case reports can be quite difficult as there is often a large volume of information which needs to be distilled and filtered to ensure only sufficiently valuable information is retained. An overly complex report will be difficult to manage and risks becoming redundant. A well-prepared safety case report should be succinct, unambiguous, accurate, focused, and defensible; in fact, the rail industry is moving towards safe case reports that occupy no more than 30 standard foolscap pages. Irrespective of the length of the report, there must be sufficient confidence in the level of detail, and accessibility of information within the report, to support a defence case in a court of law. Whilst this might seem daunting at first, there is practical advice and guidance available in various standards. For instance, BS EN 50,129 and the Yellow Book both provide advice on writing safety cases.

9.9   Structuring the Safety Case Report

There are slight differences in the way industry (BS EN 50,129) and the MOD (Def Stan 00–56) recommend the structuring of the safety case report. For instance, BS EN 50,129 suggests the following structure:

- Executive summary
- Introduction
- Definition of the system

Fig. 9.5   Testbed UK Safety
Case Framework[3]

- Quality management report
- Safety management report
- Technical safety report
- Related safety cases
- Conclusion.

In comparison, Def Stan 00–56 makes no such recommendations, nor does it define the structure of the safety case report, however it does set out three broad categories of information (Fig. 9.5):

- Information
- Reasoned arguments with a conclusion and principal assumptions
- A description of the means to be employed to prevent the identified hazards from causing accidents.

---

[3] The Safety Case Framework is a living document developed collaboratively by all the UK Testbeds, orchestrated by Zenzic. The Framework applies a consistent and transparent methodology to the development of a safety case for a CAV trial. This standardisation of risk management is a key enabler of interoperability, meaning that testing organisations can develop a safety case that is recognised by all testbeds.

Irrespective of whether the safety case report is written in accordance with BS EN 50,129 or Def Stan 00–56, a good safety report will include, as a minimum each of the following elements. Remember, the safety case report does not need to be prescriptive; merely detailed enough to provide sufficient confidence to support a defensive action in a court of law.

- *Executive summary.* The executive summary should contain a high-level description of the purpose, scope, and a summary of the main conclusions of the report and highlight any unresolved hazards and outstanding safety issues. Introduction. The introduction should clearly and succinctly set out the purpose and scope of the document and lay out the document's structure
- *System definition.* The system definition section should set the context for the safety argument; present the intended system, purpose, and functionality; define the system boundary and interfaces; present information on constituent subsystems; and provide details of the intended application
- *Quality management report.* The quality management report should provide references the quality management system(s) under which the development work has been conducted; and summarises the quality management activities carried out such as accreditation checks, supplier audits, process audits, etc.
- *Safety management report.* The safety management report is based on the safety management programme, which is presented in the safety plan. This sets out the hazard identification and analysis activities; safety requirements and targets including how these were derived and how they have been interpreted and allocated to individual subsystems; the system design review and testing process; the development of installation, operation, and maintenance procedures; the consideration and fulfilment of training needs; and any verification and validation activities
- *Technical safety report.* The technical safety report presents the technical principles which ensures the safety of the system or product, covering:
- Assurance of current functional operation
- Effects of faults
- Effects of external influences

It also demonstrates that the safety requirements and safety targets have been met (if correct); discusses any significant hazards identified and their control to reduce risks to as low as reasonably practicable; and discusses any outstanding issues with temporary control measures and proposed paths to resolution

- *Related safety cases.* When compiling related safety cases, it is important to ensure they support and corroborate the safety argument being put forward. The best way to ensure this is to ask the following question:

Does the safety argument rely on other safety cases? If so:

(1) Are they applicable to the current situation?
(2) Have any assumptions, dependencies, and caveats been complied with?

If the safety argument requires or relies on cross acceptance from other countries or applications, it is useful to consider the following issues when putting together the safety case docket:

- What is the core product and how is it applied in its native environment?
- Has the core product been accepted by a suitably accredited body and was the acceptance process consistent with the UK's regulatory regime?
- Does the product have a satisfactory performance history?
- What were the safety requirements for the core product?
- Have any differences in application and environment been rigorously identified and assessed using a risk-based approach?

These points are just a starter and are by no means comprehensive. For further advice and guidance, it may be helpful to refer to EN 50506-1:2007 Railway Applications: Communications, Signalling and Processing Systems (Application Guide for BS EN 50129 Part 1: Cross Acceptance for Guidance)

- *Conclusion.* The conclusion section to the safety case report should provide a concise summary of the whole argument. It should set out a clear statement on the acceptability of the system including whether the systems meet the safety targets and or requirements and whether the residual risk is reduced in accordance with the ALARP principle. The conclusions should also provide reference to any outstanding issues, and where they are discussed in the safety case. And finally, the conclusion should include a request for the appropriate authority to implement the system or product. The current regulatory regime for the safety of operation and installations across several industries require that a safety case is produced to demonstrate that risks associated with operating dangerous plant and machines are acceptable.

Safety cases for installations require arguments about the adequacy of individual systems within the operation where they have an impact on safety. IEC 61,508 is a complex standard which focuses on the requirements for the development of safety-related systems and the production of many documents which form part of the safety case. There is little guidance on what form an overall safety argument for the adequacy of a safety-related system should take. However, industry specific guidance is available for some industries.

Bibliography

BS EN 50129—Railway applications: Communication, signalling and processing systems (safety related electronic systems for signalling)

Corporate Manslaughter and Corporate Homicide Act 2007

Cullen Report (1988)

Def Stan 00-56—Safety Management Requirements for Defence Systems

Department for Transport (DfT—'Updating the VPF and VPIs', 2011)

DOC 178—Software Considerations in Airborne Systems and Equipment Certification

EEE-STD-1228-1994 -

Standard for Software Safety

EN 50506-1:2007 Railway Applications: Communications, Signalling and Processing Systems (Application Guide for BS EN 50129 Part 1: Cross Acceptance for Guidance)

Factories Act 1937

Factories Act 1961

Haddon-Cave Report (2009)

IEC 61508—Functional Safety of Electrical/Electronic/Programmable Electronic Safety-related Systems

JSP507—MOD Guide to Investment Appraisal and Evaluation

MIL-HDBK Electronic Reliability Design Handbook

MIL-STD-882 Department of Defence Standard Practice: System Safety (May 2012)

RAC-FMD-97 Failure Mode Distribution Data

Robens Report (1972)

United States Nuclear Regulatory Commission—Fault Tree Handbook

© The Editor(s) (if applicable) and The Author(s), under exclusive license
to Springer Nature Switzerland AG 2025
A. A. Olsen, *Hazard and Risk Analysis for Organisational Safety Management*, Synthesis
Lectures on Ocean Systems Engineering, https://doi.org/10.1007/978-3-031-73458-8